Embedded Systems

Series editors

Nikil D. Dutt, Irvine, CA, USA
Grant Martin, Santa Clara, CA, USA
Peter Marwedel, Dortmund, Germany

This Series addresses current and future challenges pertaining to embedded hardware, software, specifications and techniques. Titles in the Series cover a focused set of embedded topics relating to traditional computing devices as well as high-tech appliances used in newer, personal devices, and related topics. The material will vary by topic but in general most volumes will include fundamental material (when appropriate), methods, designs and techniques.

More information about this series at http://www.springer.com/series/8563

Djones Lettnin · Markus Winterholer
Editors

Embedded Software Verification and Debugging

Springer

Editors
Djones Lettnin
Universidade Federal de Santa Catarina
Florianópolis
Brazil

Markus Winterholer
Luzern
Switzerland

ISSN 2193-0155 ISSN 2193-0163 (electronic)
Embedded Systems
ISBN 978-1-4939-7931-8 ISBN 978-1-4614-2266-2 (eBook)
DOI 10.1007/978-1-4614-2266-2

Softcover reprint of the hardcover 1st edition 2017

Printed on acid-free paper

This Springer imprint is published by Springer Nature
The registered company is Springer Science+Business Media LLC
The registered company address is: 233 Spring Street, New York, NY 10013, U.S.A.

Markus Winterholer dedicates to
Eva-Maria and David.

Djones Lettnin dedicates to
Amelie and Fabiana.

Foreword

I am glad to write a foreword for this book.

Verification (informally defined as the process of finding bugs before they annoy or kill somebody) is an increasingly important topic. And I am particularly glad to see that the book covers the full width of verification, including debugging, dynamic and formal verification, and assertion creation.

I think that as a field matures, it goes through the following stages regarding verification:

- Trying to pay little attention to it, in an effort to "get things done";
- Then, when bugs start piling up, looking into debugging techniques;
- Then, starting to look into more systematic ways of finding new bugs;
- And finally, finding a good balance of advanced techniques, such as coverage-driven dynamic verification, improved assertions, and formal verification.

The area of HW verification (and HW/SW co-verification), where I had the pleasure of working with Markus, offers an interesting perspective: It has gone through all these stages years ago, but it was never easy to see the full path ahead.

Consider just the dynamic-verification slice of that history: Initially, no one could predict how important bugs (and thus verification) would be. It took several chip-project failures (I personally witnessed one, first hand) to understand that verification was going to be a big part of our future forever. Then, more random testing was used. That helped, but not enough, so advanced, constrained-random, massive test generation was invented. Then, it became clear that functional coverage (not just code coverage) was needed, to make sense of all the resulting runs and see which covered what.

It then dawned on everybody that this new coverage-driven verification needed its own professionals, and thus "verification engineer" as a job description came to be. Then, as CDV started producing more failing runs than engineers could debug, emphasis again shifted to advanced debug tools and so on. All of this looks reasonable in hindsight, but was not so obvious on day one.

Newer fields like autonomous systems are still working their way through the list, but as the cost of bugs there become clearer, I expect them to adopt more of the advanced techniques mentioned in this book.

November 2016

Yoav Hollander
Foretellix Ltd.

Contents

Contributors

Ziad A. Al-Sharif Software Engineering Department, Jordan University of Science and Technology, Irbid, Jordan

Raimundo Barreto Federal University of Amazonas, Manaus, Brazil

Edna Barros CIn - Informatics Center, UFPE—Federal University of Pernambuco, Recife, Brazil

Jörg Behrend Department of Computer Engineering, University of Tübingen, Tübingen, Germany

Lucas Cordeiro Federal University of Amazonas, Manaus, Brazil

Hanno Eichelberger University of Tübingen, Tübingen, Germany

Alexander Grünhage Department of Computer Engineering, University of Tübingen, Tübingen, Germany

Hussama Ismail Federal University of Amazonas, Manaus, Brazil

Padma Iyenghar Software Engineering Research Group, University of Osnabrueck, Osnabrück, Germany

Clinton L. Jeffery Computer Science Department, University of Idaho, Moscow, ID, USA

Jan Kotas Cadence® Design Systems, Katowice, Poland

Thomas Kropf Department of Computer Engineering, University of Tübingen, Tübingen, Germany

Djones Lettnin Department of Electrical and Electronic Engineering, Federal University of Santa Catarina, Trindade, Florianópolis, SC, Brazil

Rafael Melo Macieira CIn - Informatics Center, UFPE—Federal University of Pernambuco, Recife, Brazil

Elke Pulvermueller Software Engineering Research Group, University of Osnabrueck, Osnabrück, Germany

Herbert Rocha Federal University of Roraima, Boa Vista, Brazil

Wolfgang Rosenstiel Department of Computer Engineering, University of Tübingen, Tübingen, Germany

Jürgen Ruf Department of Computer Engineering, University of Tübingen, Tübingen, Germany

Mahmoud H. Said Software Engineering Department, Jordan University of Science and Technology, Irbid, Jordan

Michael Uelschen University of Applied Sciences, Osnabrück, Germany

Clemens Westerkamp University of Applied Sciences, Osnabrück, Germany

Markus Winterholer swissverified.com, Lucerne, Switzerland

Cyprian Wronka Cadence® Design Systems, San Jose, CA, USA

Juergen Wuebbelmann University of Applied Sciences, Osnabrück, Germany

Chapter 1
An Overview About Debugging and Verification Techniques for Embedded Software

Djones Lettnin and Markus Winterholer

1.1 The Importance of Debugging and Verification Processes

Embedded systems (ES) have frequently been used over the last years in the electronic systems industry due to their flexible operation and possibility of future expansions. Embedded systems are composed of hardware, software, and other modules (e.g., mechanics) designed to perform a specific task as part of a larger system. Important further concepts such as Cyber-Physical Systems (CPS) and Internet of Things (IoT) consider also different aspects of ES. In CPS, computation and physical processes are integrated considering physical quantities such as timing, energy, and size [4]. In IoT, physical objects are seamlessly integrated into the information network [47]. Taking everything into account, internal control of vehicles, autopilot, telecommunication products, electrical appliances, mobile devices, robot control, and medical devices are some of the practical examples of embedded systems.

Over the last years, the amount of software used in embedded electronic products has been increasing and the tendency is that this evolution continues in the future. Almost 90% of the microprocessors developed worldwide have been applied in embedded systems products [52], since the embedded software (ESW) is the main responsible for functional innovations, for instance, in the automotive area with the reduction of gas emissions or with the improvement of security and comfort [45].

The embedded software is also frequently used in safety critical applications (e.g., automotive) where failures are unacceptable [21], as seen in lists of disasters

D. Lettnin (✉)
Department of Electrical and Electronic Engineering, Federal University of Santa Catarina, Florianópolis, Brazil
e-mail: djones.lettnin@ufsc.br

M. Winterholer
swissverified.com, Lucerne, Switzerland
e-mail: markus@winterholer.com

D. Lettnin and M. Winterholer (eds.), *Embedded Software Verification and Debugging*, Embedded Systems, DOI 10.1007/978-1-4614-2266-2_1

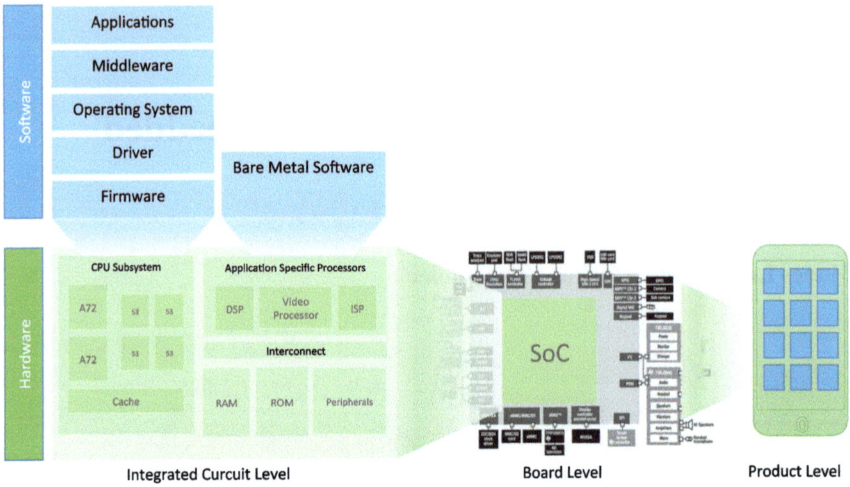

Fig. 1.1 Example a SoC into a system

and inconveniences occurred due to software errors [26, 32]. The main challenge of verification and debugging processes is to handle the system complexity. For instance, the automotive embedded software of a car achieved up to 1 GB by 2010 [61]. As it can be observed in Fig. 1.1, embedded software is being applied with different views in modern SoCs, going from application software (e.g., apps, middleware, operating system, drivers, firmware) distributed among many processor cores, as well as, hardware-dependent (i.e., bare metal) software and finally, covering the communication software stacks.

The electronic system level (ESL) design and verification consider usually a combination of bottom-up and top-down approaches [63]. It meets the system-level objectives by exploiting the synergism of hardware and software through their concurrent design. Therefore, the development of software needs start earlier (in parallel) to the SoC design, integration‘, and verification, as depicted in Fig. 1.2. During the pre-silicon phase, it is time to remove critical bugs in system environment. In this phase, the SW is becoming more and more a requirement to tape out, since it may hold the fabrication if a bug is too critical. After the production, the development of SW can be continued on-chip and the post-silicon validation will be performed.

Software development, debugging, and verification processes are driving SoC project costs reaching up to 80% of overall development costs, as it can be observed in Fig. 1.3. The design complexity is getting higher and for this reasons it originates the design productivity gap and the verification gap. The technology capability is currently doubling every 36 months. The hardware design productivity improved over the last couple of years by filling the silicon with multi-core and with memory components, and providing additional functionality in software [42]. With the increase amount of embedded software, a software gap could be noticed, where the main challenge now is how to fit millions of software lines with millions of

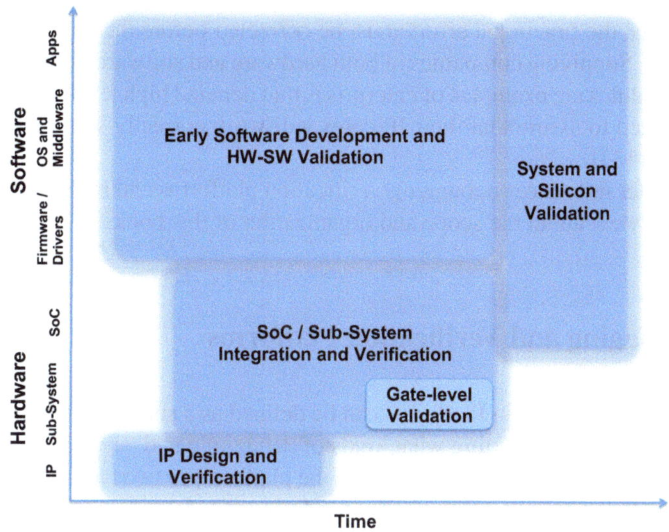

Fig. 1.2 System development. Adapted from [37]

Fig. 1.3 Software and verification driving SoC project costs [37]

gates [10]. The software part is currently doubling every 10 months, however, the productivity for hardware-dependent software only doubles every 5 years. Together with the increase of the design complexity, the lifetime and the time-to-market requirements have been demanding shorter system design periods. This development period could be smaller if it would be possible to minimize the verification and debugging time [68]. When a device needs to be re-designed and/or new project cycles need to be added to the development due to design errors, the final cost of the product can be increased by hundreds of thousands of dollars. It is also common agreement that the functional errors must be corrected before the device is released to the market. Supplying companies of both hardware and software intellectual property (IP[1]) modules are examples of enterprises that demand high level of correctness, since they need to assure that their IP cores will work correctly when inserted in a target project [33].

This chapter introduces debugging/verification platforms and methodologies and gives an overview about the scope and organization of this book.

1.2 Debugging and Verification Platforms

Debugging and Verification Platforms can be defined as a standard for the hardware of a computer system, deciding what kinds of debugging and verification processes can be performed. Basically, we can divide the platforms in two categories: Pre- and Post-Silicon. In the pre-silicon platforms, the designs are debugged and verified using virtual environment with sophisticated simulation and formal verification tools. In distinction to post-silicon platforms where real devices are used running on target boards with logic analyzer and assertion-based tools.

1.2.1 OS Simulation

The operating systems of smart devices (e.g., smartphones) allow the developers to create thousands of additional programs with several utilities, such as, to store personal data of the users. In order to develop these applications (i.e., apps), each platform has its strengths, weaknesses, and challenges.

Gronli et al. [36] compare the main mobile OS platforms in several different categories, such as software architecture, application development, platform capabilities and constraints, and, finally, developer support. The compared OS platforms considers: (1) Android, a Linux-based operating system from Google; (2) The Windows Phone operating system from Microsoft; (3) The iOS platform from Apple; and one platform representing a new generation: (4) The new web-based Firefox OS from Mozilla. All evaluated platforms presented from good to excellent interactive debugging options.

[1]Intellectual property cores are design modules of both hardware or software units used as building blocks, for instance, within SoC designs.

1.2.2 Virtual Platform

Virtual prototyping (VP) [3, 6] (aka. Virtual Platforms) is a software model of a system that can be used for early software development and SoC/system architectural analysis. Virtual prototyping includes processor and peripheral models that may run at or near real-time speeds, potentially supporting pre-silicon development of the entire software stack up to the applications level. Virtual prototyping solutions may come with tools to help develop models, and they also usually provide a simulation and debugging environment [72]. While early virtual prototyping solutions required proprietary models, many now use SystemC models based on the Open SystemC Initiative (OSCI), transaction-level modeling (TLM), [7] standard and the IEEE-1666 SystemC standard [58].

In addition to early software development, virtual prototyping can be used for software distribution, system development kits and customer demos. In post-RTL software development, for example, virtual prototyping can be used as a low-cost replacement for silicon reference boards distributed by semiconductor companies to software developers in systems companies. Compared to reference boards, virtual prototyping provides much better debug capabilities and iteration time, and therefore can accelerate the post-silicon system integration process [6].

1.2.3 RTL Simulation

Hardware-dependent software requires a simulator or a target platform to be tested. Register Transfer Level (RTL) simulation is the most widely used method to validate the correctness of digital IC designs. They are better suited to test software with hardware dependencies (e.g., assembly code) and that requires timing accuracy. However, when simulating a large IC designs with complicated internal behaviors (e.g., CPU cores running embedded software), RTL simulation can be extremely time consuming. Since RTL-to-layout is still the most prevalent IC design methodology, it is essential to speedup the RTL simulation process. Recently, General Purpose computing on Graphics Processing Units (GPGPU) is becoming a promising paradigm to accelerate computing-intensive workloads [62].

1.2.4 Acceleration/Emulation

Traditional debugging tools have not kept pace with the rapid rate at which system-on-chip (SoC)/ASIC design size and complexity are growing. As RTL/gate design size increases, traditional simulators slowdown significantly, which delays hardware/software (system) integration and prolong the overall verification cycle.

When excessive simulation time becomes a bottleneck for dynamic verification, hardware emulation and simulation acceleration are often used. Hardware emulators provide a debugging environment with many features that can be found in logic simulators, and in some cases even surpass their debugging capabilities, such as setting breakpoints and visibility of content or sign in memory design. For the Assertion-based Verification (ABV) methodology to be used in hardware emulation, assertions must be supported in hardware [12]. Traditional emulators are based on reconfigurable logic and FPGAs. To increase flexibility and to ease the debugging process, which requires the ability to instrument assertions, current-generation emulators and simulation accelerators are typically based on an array of processing elements, such as in Cadence Palladium [15]. Another approach, is to integrate the debug and communication module inside the chip such as an on-chip in-circuit emulation (ICE) architecture for debugging [75]. However, due to its high cost, emulators are expensive for many developers.

1.2.5 FPGA Prototyping

During the last years, Commercial-Off-The-Shelf (COTS) FPGAs provide processing capability fulfilling the demand required by the increasing instruments resolution and measurement speeds, even with low power budget [55]. Furthermore, partial dynamic reconfiguration permits changing or adapting payload processing during operation.

FPGA technology is commonly used to prototype new digital designs before entering fabrication. Whilst these physical prototypes can operate many orders of magnitude faster than through a logic simulator, a fundamental limitation is their lack of on-chip visibility when debugging. In [41] a trace-buffer-based instrumentation was installed into the prototype, allowing designers to capture a predetermined window of signal data during live operation for offline analysis. However, instead of requiring the designer to recompile their entire circuit every time the window is modified, it was proposed that an overlay network is constructed using only spare FPGA routing multiplexers to connect all circuit signals through to the trace instruments. Thus, during debugging, designers would only need to reconfigure this network instead of finding a new place-and-route solution.

1.2.6 Prototyping Board

Traditionally, post-silicon debugging is usually painful and slow. Observability into silicon is limited and is expensive to achieve. Simulation and emulation is slow and is extremely tough to hit corner-case scenarios, concurrent and cycle-dependent behavior. With simulation , the hope is that the constrained-random generator will hit the input combination, which caused the failure scenario (triggered the bug). Not

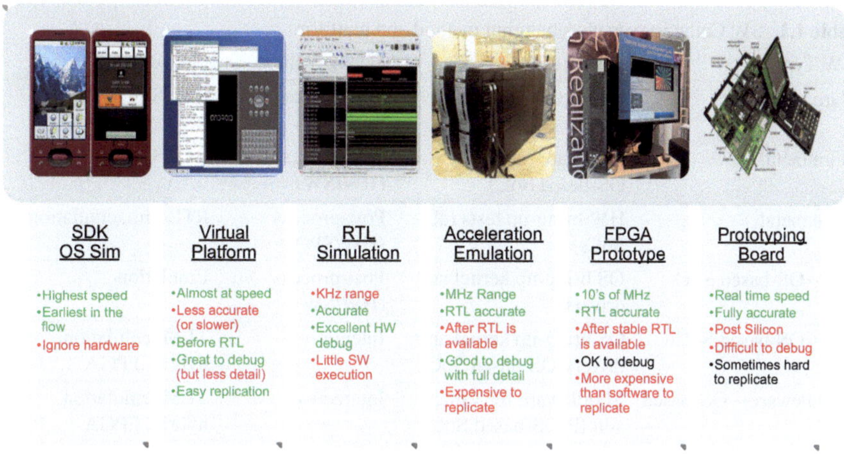

Fig. 1.4 Strength and weakness of each platform [44]

the least, time-to-market is a major concern when complex post-silicon bugs surface, and it takes time to find the root cause and the fix of the issue [5].

Post-silicon introspection techniques have emerged as a powerful tool to combat increased correctness, reliability, and yield concerns. Previous efforts using post-silicon introspection include online bug detection, multiprocessor cache-coherence validation, online defect detection. In [22] an Access-Control Extensions (ACE) was proposed that can access and control a microprocessor's internal state. Using ACE technology, special firmware can periodically probe the microprocessor during execution to locate run-time faults, repair design errors.

1.2.7 Choosing the Right Platform for Software Development and Debugging

As it could be observed in the previous sections, there is no "one fits all" approach. Each platform has strength and weakness, as can be summarized in Fig. 1.4.

1.3 Debugging Methodologies

Debugging might be considered as the most time-consuming task for embedded system developers. Any investment in effective debug tools and infrastructure accelerates the product release. Basically the debugging process can be performed in Interactive or Post-Process forms.

Table 1.1 SW Category versus debugging method and platform

SW Type	SW category	Debug method	Platforms
Bare metal	Boot ROM (all SoCs)	Post-process (HW/SW)	RTL Sim, emulation
Bare metal	Firmware (non OS-based SoC)	Post-process (HW/SW)	RTL Sim, emulation
Bare metal	HW bring-up tests (all SoCs)	Post-process (HW/SW)	RTL Sim, emulation
OS—OS-based SoC	OS bring-up kernel and drivers	Post-process (HW/SW)	Emulation
OS—OS-based SoC	OS bring-up kernel and drivers(OS-based SoC)	Interactive	TLM, emulation, hybrid, FPGA
Middleware—OS-based SoC	Middleware for value add IP(OS-based SoC)	Interactive	TLM, emulation, hybrid, FPGA
Application—OS-based SOC	Application tests for value add IP (OS-based SoC)	Interactive	TLM, emulation, hybrid, FPGA

1.3.1 Interactive Debugging

One popular debug approach that overcomes the observability problem is the so-called interactive (or run/stop) technique, which stops an execution of the SoC before its state is inspected in detail. An advantage of this technique is that we can inspect the SoC's full state without running into the device pins' speed limitations. It also requires only a small amount of additional debug logic in the SoC. The main disadvantage of interactive debug is that the technique is intrusive, since the SoC must be stopped prior to observing its state [71].

The most primitive forms of debugging are the printing of messages on the standard output (e.g., *printf* of C language) and the usage of debugging applications (e.g., *gdb*). If the embedded software is being tested on a hardware engine, *JTAG*[2] interfaces should be used to acquire debugging information [2]. As example of industrial debug solutions are: Synopsys System-Level Catalyst [46, 69] with focus on virtual platforms and FPGA prototypes debugging; SVEN and OMAR focuses on software and hardware technologies increasing silicon and software debug facilities [13].

1.3.2 Post-Process Debugging

The Post-Process Debugging runs simulation in batch mode, record data to a waveform database, and then analyze the results after simulation is complete. This latter

[2]JTAG (Joint Test Action Group) is a hardware interface based on the IEEE 1149.1 standard and used for scan testing of printed circuit boards.

Table 1.2 Embedded software debug

	JTAG	TLM Sim	Interactive debug: RTL Sim	Post-process debug: RTL Sim	Interactive debug: Emulation	Post-process debug: emulation	Interactive debug: emulation hybrid	Interactive debug: simulation hybrid
SW execution speed	Good	Good	Poor	NA	Emulation speed	NA	Implementation dependent	Implementation dependent
SW execution timing accuracy	Good	Poor	Good	Good	Good	Good	Poor	Poor
SW visibility (stack, memory, variables, etc.)	Good	Good	Implementation dependent	Implementation dependent	Implementation dependent	Implementation dependent	Good	Good
Intrusive SW debug	Yes	No	Implementation dependent	No	Implementation dependent	No	Implementation dependent	Implementation dependent
HW visibility	Programmer view	Good	Good	Everything traced	Emulator visibility	Emulator visibility	Emulator visibility	Good
HW accuracy	Good	Poor	Good	Good for traced HW	Good	Good for traced HW	Good	Good
HW/SW synchronization	Programmer view	Good	Implementation dependent	Good	Good	Good	Implementation dependent	Implementation dependent
Impact of Post-process data collection for SW	NA	NA	NA	Implementation dependent	NA	Implementation dependent	NA	NA
Impact of Post-process data collection for HW	NA	NA	NA	Implementation dependent	NA	Implementation dependent	NA	NA

use model presents challenges when trying to post-process results when the DUT is being driven by a class-based verification environment, such as the Open Verification Methodology (OVM) or Universal Verification Methodology (UVM) [18].

The Incisive Debug Analyzer (IDA) [16] provides functionality of an interactive debug flow plus the advantage of debugging in post-process mode, allowing all the debug data files running the simulation once.

1.3.3 Choosing the Right Debugging Methodology

Table 1.1 correlates the SW category and the debugging methods as well as the debugging platforms.

As it could be observed in the previous sections, both debugging methods have their strength and weakness, as can be summarized in Table 1.2.

1.4 Verification Methodologies

1.4.1 Verification Planning

Verification planning is a methodology that defines how to measure variables, scenarios, and features. Additionally, it documents how verification results are measured considering, for instance, simulation coverage, directed tests, and formal analysis. It also provides a framework to reach consensus and to define verification closure for a design. An example of verification planning tool is the Enterprise Planner [17], which allows to create, edit, and maintain verification plans, either starting from scratch, or by linking and tracking the functional specifications.

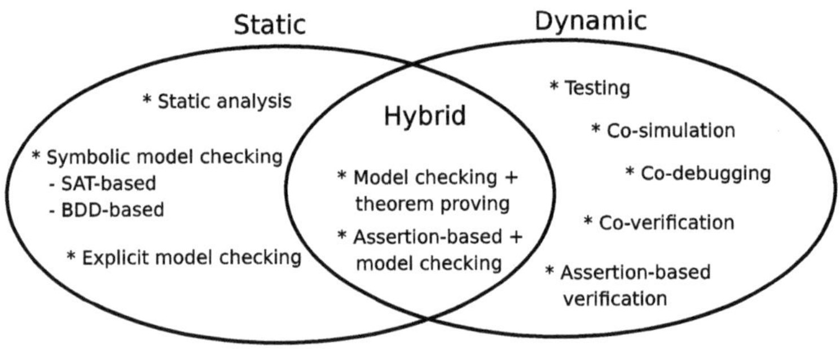

Fig. 1.5 Taxonomy of embedded systems verification approaches [48]

1.4.2 *Verification Environment Development*

Figure 1.5 illustrates the taxonomy of embedded system verification approaches. Dynamic verification needs to execute the embedded system during the verification process. It focuses mainly on testing, co-verification, and assertion-based verification. On the other hand, static verification verifies the embedded system without its execution. It is presented with focus on static analysis and model checking approaches. Theorem proving demands skilled user iteration and is mainly used in combination with other static verification approaches. Finally, hybrid approaches are focused on the combination of static approaches and of dynamic-static approaches [48].

1.4.2.1 Dynamic Verification

The dynamic verification focuses on testing, co-verification, and assertion-based verification approaches. Dynamic verification for hardware-independent software can be tested directly on a host machine. On the other hand, hardware-dependent software requires a simulator or a target platform. If the embedded software demands high performance (e.g., operating system booting, video processing applications) a hardware engine (e.g., in-circuit-emulator, simulation accelerator, or rapid prototyping) can be used to increase performance. The main advantage of dynamic verification is that the whole system can be used in the verification in order to test more deeply into the system state space.

Testing

Testing is an empirical approach that intent to execute the software design in order to identify any design errors [8]. If the embedded software does not work, it should be modified in order to get it work. Scripting languages are used for writing different test scenarios (e.g., functions with different parameter values or different function call sequences). The main testing methodologies and techniques are listed in the following [67, 74]:

Metric-driven Hardware/Software Co-verification

Metric-driven verification is the use of a verification plan and coverage metrics to organize and manage the verification project, and optimize daily activities to reach verification closure. Testbenches are designed in order to drive inputs into hardware/software modules and to monitor internal states (white box verification[3]) or the output results (black box verification[4]) of the design. Executing regression suites produces a list of failing runs that typically represent bugs in the system to resolve, and coverage provides a measure of verification completion. Bugs are iteratively fixed, but the unique process of metric-driven verification is the use of

[3]White box verification focus on knowledge of a system's internal structure [8].

[4]Black box verification focus on the functional behavior of the system, without explicit knowledge of the internal details [8].

coverage charts and coverage hole analysis to aid verification closure. Analyzing coverage holes provides insight into system scenarios that have not been generated, enabling the verification team to make adjustments to the verification environment to achieve more functional coverage [14].

As an example, coverage driven verification has been successfully used in the hardware area with the ***e*** language. Recently, it has been extended to embedded software through the Incisive Software extensions (ISX) [73].

Assertion-Based Verification

Assertion-based verification methodology captures a design's intended behavior in temporal properties and monitors the properties during system simulation [30]. After the specification of system requirement, the informal specification is cast into temporal properties that capture the design intent. This formalization of the requirements already improves the understanding of the new system. This methodology has been successfully used at lower levels of hardware designs, specially at register transfer level (RTL), which requires a clock mechanism as timing reference and signals at the Boolean level [30]. Thus, it is not suitable to apply this hardware verification technique directly to embedded software, which has no timing reference and contains more complex structures (e.g., integers, pointers, etc.). Thus, new mechanisms are used in order to apply assertion-based methodology with embedded software [50, 51].

1.4.2.2 Static Verification

Static verification performs analysis without the execution of the program. The analysis is performed on source or on object code. Static verification of embedded software focuses mainly on abstract static analysis, model checking and theorem proving.

Static Analysis

Static analysis has been widely used in the optimization of compiler design (e.g., pointer analysis). In the software verification, static analysis has been used for highlighting possible coding errors (e.g., linting tools) or formal static analysis in order to verify invariant properties, such as division-by-zero, array bounds, and type casting [25, 53]. This approach has been also used for the analysis of worst case execution time (WCET) and of stack/heap memory consumption [1].

Formal static analysis is based on abstract interpretation theory [24], which approximates to the semantics of program execution. This approximation is achieved by means of abstract functions (e.g., numerical abstraction or shape analysis) that are responsible for mapping the real values to abstract values. This model over-approximates the behavior of the system to make it simple to analyze. On the other hand, it is incomplete and not all real properties of the original system are valid for the abstract model. However, if the property is valid in abstract interpretation then the property is also valid in the original system.

Model Checking

Model checking (MC) verifies whether the model of the design satisfies a given specification. There are two main paradigms for model checking: Explicit state model checking and symbolic model checking.

Explicit state model checking uses an explicit representation (e.g., hash table) in order to store the explored states given by a state transition function. On the other hand, symbolic model checking [21] stores the explored states symbolically in a compact form. Although explicit model checking has been used in the software verification (e.g., SPIN model checker [40]), it has memory limitations compared with symbolic model checkers [60]. The symbolic model checking is based on binary decision diagrams (BDD) [54] or on Boolean satisfiability (SAT) [31] and it has been applied in the formal verification process. However, each approach has its own strengths and weaknesses [59].

Formal verification can handle up to medium-sized software systems, where they have less state space to explore. For larger software designs, formal verification using model checking often suffers from the state space explosion problem. Therefore, abstraction techniques are applied in order to alleviate the burden for the back-end model checkers.

The commonly software model checking approaches are as follows:

- Convert the C program to a model and feed it into a model checker [43].
  This approach models the semantics of programs as finite state systems by using suitable abstractions. These abstract models are verified using both BDD-based and SAT-based model checkers.
- Bounded Model Checking (BMC) [19].
  This approach unwinds the loops in the embedded software and the resulting clause formula is applied to a SAT-based model checker.

Each approach has its own strengths and weaknesses and a detailed survey on software model checking approaches is made in [29].

The main weaknesses of current model checking approaches are still the modeling of suitable abstraction models and the state space explosion for large industrial embedded software.

Theorem Proving

Theorem proving is a deductive verification approach, which uses a set of axioms and a set of inference rules in order to prove a design against a property. Over the last years, the research in the field of automated theorem provers (ATP) has been an important topic in the embedded software verification [27]. However, the standalone theorem proving technique still needs skilled human guidance in order to construct a proof.

1.4.2.3 Hybrid Verification

The combination of verification techniques is an interesting approach in order to overcome the drawbacks of the isolated dynamic and static aforementioned verification approaches.

The main hybrid verification approach for the verification of embedded software has been focused on combining model checking and theorem proving, such as satisfiability modulo theories (SMT) [66] and predicate abstraction approaches.

SMT combines theories (e.g., linear inequality theory, array theory, list structure theory, bit vector theory) expressed in classical first-order logic in order to determine if a formula is satisfiable. The predicate symbols in the formula may have additional interpretations that are classified according to the theory that they belong to. In this sense, SMT has the advantage that a problem does not have to be translated to Boolean level (like in SAT solving) and can be handled on word level. For instance, SMT-based BMC has been used in the verification of multi-threaded software allowing the state space to be reduced by abstracting the number of state variables and interleavings from the proof of unsatisfiability generated by the SMT solvers [23].

Model checking with predicate abstraction using a theorem prover [11] or a SAT-solver [20] checks software based on an *abstract-check-refine* paradigm. It constructs an abstract model based on predicates and then checks the safety property. If the check fails, it refines the model and iterates the whole process.

The combination of dynamic and static verification has been explored in the hardware verification domain [28, 34, 35, 38, 39, 56, 57, 64, 65, 70]. Basically, simulation is used to reach "interesting" (also known as lighthouse or critical) states. From these states, the model checker can verify exhaustively a local state space for a certain number of time steps. This approach is available in the hardware commercial verification tools such as Magellan [39]. Additionally, the combination of simulation and formal verification has been applied to find bugs in the verification of hardware serial protocols, in which isolated techniques (i.e., only simulation or only formal verification) were unable to find them [34].

One way to control the embedded software complexity lies in the combination of formal methods with simulative approaches. This approach combines the benefits of going deep into the system and covers exhaustively the state space of the embedded software system. For example, assertion-based verification and formal verification based on state-of-the-art software model checkers are combined and applied to the verification of embedded software using the C language [9, 48, 49].

1.5 Summary

This chapter has presented and discussed the main merits and shortcomings of the state-of-the-art in debugging and verification of embedded software.

References

1. AbsInt: WCET analysis and stack usage analysis. http://www.absint.com/
2. Andrews J (2005) Co-verification of hardware and software for ARM SoC Design. Newnes
3. Andrews J (2011) Welcome to the cadence virtual system platform. URL http://www.cadence.com/
4. Ashton K (2009) That 'internet of things' thing, in the real world things matter more than ideas. RFID 1:10–10
5. Automation JD, Post-silicon debug solution. http://www.jasper-da.com/products/post-silicon-debug
6. Avinun R (2011) Concurrent hardware/software development platforms speed system integration and bring-up. http://www.cadence.com/
7. Bailey B, McNamara M, Balarin F, Stellfox M, Mosenson G, Watanabe Y (2010) TLM-driven design and verification methodology. Cadence Des Syst
8. Bart B, Noteboom E (2002) Testing embedded software. Addison-Wesley Longman
9. Behrend J, Lettnin D, Heckeler P, Ruf J, Kropf T, Rosenstiel W (2011)Scalable hybrid verification for embedded software. In: DATE '11: Proceedings of the conference on design, automation and test in Europe, pp 179–184
10. Berthet C (2002) Going mobile: the next horizon for multi-million gate designs in the semiconductor industry. In: DAC '02: Proceedings of the 39th conference on Design automation, ACM, New York, USA, pp 375–378. doi:10.1145/513918.514015
11. Beyer D, Henzinger TA, Jhala R, Majumdar R (2007) The software model checker BLAST. STTT 9(5–6):505–525
12. Boul M, Zilic Z (2008) Generating hardware assertion checkers: for hardware verification, emulation, post-fabrication debugging and on-line monitoring, 1st edn. Springer, Incorporated
13. Brouillette P (2010) Accelerating soc platform software debug with intelÕs sven and omar. In: System, software, SoC and silicon debug S4D conference 2010
14. Brown S (2011) Hardware/software verification with incisive software extensions. http://www.cadence.com/
15. Cadence design systems: cadence palladium. http://www.cadence.com
16. Cadence design systems: incisive debug analyzer. http://www.cadence.com
17. Cadence design systems: incisive management. http://www.cadence.com
18. Cadence design systems: post-processing your ovm/uvm simulation results. http://www.cadence.com
19. Clarke E, Kroening D, Lerda F (2004) A tool for checking ANSI-C programs. In: Jensen K, Podelski A (eds) TACAS: tools and algorithms for the construction and analysis of systems (TACAS 2004), Lecture notes in computer science, vol 2988. Springer, pp 168–176
20. Clarke E, Kroening D, Sharygina N, Yorav K (2005) SATABS: SAT-based predicate abstraction for ANSI-C. In: TACAS:Tools and Algorithms for the Construction and Analysis of Systems (TACAS 2005), Lecture notes in computer science, vol 3440, pp 570–574. Springer
21. Clarke EM, Grumberg O, Peled DA (1999) Model checking. The MIT Press
22. Constantinides K, Austin T (2010) Using introspective software-based testing for post-silicon debug and repair. In: Design automation conference (DAC), 2010 47th ACM/IEEE, pp 537–542
23. Cordeiro L (2010) Smt-based bounded model checking for multi-threaded software in embedded systems. In: Proceedings of the 32nd ACM/IEEE international conference on software engineering-volume 2, ICSE 2010, Cape Town, South Africa, 1–8 May 2010, pp 373–376
24. Cousot P, Cousot R (1977) Abstract interpretation: a unified lattice model for static analysis of programs by construction or approximation of fixpoints. In: Conference record of the fourth annual ACM SIGPLAN-SIGACT symposium on principles of programming languages, ACM Press, New York, pp 238–252
25. Coverity: Coverity static analysis verification engine (coverity save). http://www.coverity.com/products/coverity-save/
26. Dershowitz N, The software horror stories. http://www.cs.tau.ac.il/~nachumd/horror.html

27. Detlefs D, Nelson G, Saxe JB (2003) Simplify: a theorem prover for program checking. Tech Rep J ACM
28. Dill DL, Tasiran S (1999) Formal verification meets simulation. In: ICCAD '99: Proceedings of the 1999 IEEE/ACM international conference on Computer-aided design, IEEE Press, Piscataway, NJ, USA. Chairman-Ellen M. Sentovich, p 221
29. D'Silva V, Kroening D, Weissenbacher G (2008) A survey of automated techniques for formal software verification. TCAD: IEEE Trans Comput Aided Des Integr Circ Syst 27(7):1165–1178. doi:10.1109/TCAD.2008.923410
30. Foster HC, Krolnik AC, Lacey DJ (2004) Assertion-based design. Springer
31. Ganai M, Gupta A (2007) SAT-based scalable formal verification solutions. Springer
32. Ganssle J (2006) Total recall. http://www.embedded.com/
33. Goldstein H (2002) Checking the play in plug-and-play. IEEE Spectr 39:50–55
34. Gorai S, Biswas S, Bhatia L, Tiwari P, Mitra RS (2006) Directed-simulation assisted formal verification of serial protocol and bridge. In: DAC '06: proceedings of the 43rd annual conference on Design automation, ACM Press, New York, USA, pp 731–736. doi:10.1145/1146909.1147096
35. Gott RM, Baumgartner JR, Roessler P, Joe SI (2005) Functional formal verification on designs of pseries microprocessors and communication subsystems. IBM J. 49(4/5):565–580
36. Grønli TM, Hansen J, Ghinea G, Younas M (2014) Mobile application platform heterogeneity: Android vs windows phone vs ios vs firefox os. In: Proceedings of the 2014 IEEE 28th international conference on advanced information networking and applications, AINA '14, IEEE Computer Society, Washington, DC, USA, pp 635–641. doi:10.1109/AINA.2014.78
37. Hanna Z (2014) Challenging problems in industrial formal verification. In: Proceedings of the 14th conference on formal methods in computer-aided design, FMCAD '14, FMCAD Inc, Austin, TX. pp 1:1–1:1. http://dl.acm.org/citation.cfm?id=2682923.2682925
38. Hazelhurst S, Weissberg O, Kamhi G, Fix L (2002) A hybrid verification approach: getting deep into the design
39. Ho PH, Shiple T, Harer K, Kukula J, Damiano R, Bertacco V, Taylor J, Long J (2000) Smart simulation using collaborative formal and simulation engines. ICCAD 00:120. doi:10.1109/ICCAD.2000.896461
40. Holzmann GJ (2004) The spin model checker: primer and reference manual. Addison-Wesley
41. Hung E, Wilton SJE (2014) Accelerating fpga debug: increasing visibility using a runtime reconfigurable observation and triggering network. ACM Trans. Des. Autom. Electron. Syst. 19(2), 14:1–14:23. doi:10.1145/2566668
42. ITRS: International technology roadmap for semiconductors (2007). http://www.itrs.net/
43. Ivanicic F, Shlyakhter I, Gupta A, Ganai MK (2005) Model checking C programs using F-SOFT. In: ICCD '05: proceedings of the 2005 international conference on computer design, IEEE Computer Society, Washington, DC, USA, pp 297–308. doi:10.1109/ICCD.2005.77
44. Kevan T, Managing complexity with hardware emulation. http://electronics360.globalspec.com/article/4336/managing-complexity-with-hardware-emulation
45. Kropf T (2007) Software bugs seen from an industrial perspective or can formal method help on automotive software development?
46. Lauterbach S, Trace32. http://www.lauterbach.com/
47. Lee EA (2007) computing foundations and practice for cyber-physical systems: a preliminary report. Tech Rep UCB/EECS-2007-72
48. Lettnin D (2010) Verification of temporal properties in embedded software: based on assertion and semiformal verification approaches. Suedwestdeutscher Verlag fuer Hochschulschriften
49. Lettnin D, Nalla PK, Behrend J, Ruf J, Gerlach J, Kropf T, Rosenstiel W, Schönknecht V, Reitemeyer S (2009) Semiformal verification of temporal properties in automotive hardware dependent software. In: DATE '09: proceedings of the conference on design, automation and test in Europe
50. Lettnin D, Nalla PK, Ruf J, Kropf T, Rosenstiel W, Kirsten T, Schönknecht V, Reitemeyer S (2008) Verification of temporal properties in automotive embedded software. In: DATE '08: Proceedings of the conference on Design, automation and test in Europe, ACM, New York, NY, USA, pp 164–169. doi:10.1145/1403375.1403417

51. Lettnin D, Rosenstiel W (2008) Verification of temporal properties in embedded software. In: IP 08: IP-based system design, Grenoble
52. Liggesmeyer P, Rombach D (2005) Software-engineering eingebetteter systeme: Grundlagen-Methodik-Anwendungen, vol 1. Spektrum Akademischer Verlag
53. MathWorks T, Polyspace embedded software verification. http://www.mathworks.com/products/polyspace/
54. McMillan KL (1992) Symbolic model checking: an approach to the state explosion problem. Ph.D thesis, Pittsburgh, PA, USA
55. Michel H, Bubenhagen F, Fiethe B, Michalik H, Osterloh B, Sullivan W, Wishart A, Ilstad J, Habinc S (2011) Amba to socwire network on chip bridge as a backbone for a dynamic reconfigurable processing unit. In: 2011 NASA/ESA Conference on Adaptive hardware and systems (AHS), pp 227–233. doi:10.1109/AHS.2011.5963941
56. Mony H, Baumgartner J, Paruthi V, Kanzelman R, Kuehlmann A (2004) Scalable automated verification via expert-system guided transformations. http://citeseer.ist.psu.edu/mony04scalable.html
57. Nanshi K, Somenzi F (2006) Guiding simulation with increasingly refined abstract traces. In: DAC '06: Proceedings of the 43rd annual conference on design automation, ACM Press, NY, USA, pp 737–742. doi:10.1145/1146909.1147097
58. (OSCI), O.S.I.: IEEE 1666 standard systemc language reference manual (LRM) (2005)
59. Parthasarathy G, Iyer MK, Cheng KT (2003) A comparison of BDDs, BMC, and sequential SAT for model checking. In: HLDVT '03: Proceedings of the eighth IEEE international workshop on high-level design validation and test workshop, IEEE Computer Society, Washington, DC, USA, p 157
60. Peled D (2002) Comparing symbolic and explicit model checking of a software system. In: In Proceedings of SPin workshop on model checking of software, vol 2318. LNCS, Springer, pp 230–239
61. Pretschner A, Broy M, Kruger IH, Stauner T (2007) Software engineering for automotive systems: a roadmap. In: FOSE '07: 2007 future of software engineering, IEEE Computer Society, Washington, DC, USA, pp 55–71. doi:10.1109/FOSE.2007.22
62. Qian H, Deng Y (2011) Accelerating rtl simulation with gpus. In: IEEE/ACM international conference on computer-aided design (ICCAD), 2011, pp 687–693. doi:10.1109/ICCAD.2011.6105404
63. Rigo S, Azevedo R, Santos L (2011) Electronic system level design: an open-source approach. Springer
64. Ruf J, Kropf T (2002) Combination of simulation and formal verification. In: Proceedings of GI/ITG/GMM-workshop Methoden und Beschreibungssprachen zur Modellierung und Verifikation von Schaltungen und Systemen. Shaker Verlag
65. Shyam S, Bertacco V (2006) Distance-guided hybrid verification with GUIDO. In: DATE '06: Proceedings of the conference on design, automation and test in Europe, European design and automation association, 3001 Leuven, Belgium, pp 1211–1216
66. SMT-Exec: Satisfiability modulo theories execution service. http://www.smtcomp.org/
67. Spillner A, Linz T, Schaefer H (2006) Software testing foundations: a study guide for the certified tester exam. O'Reilly media
68. Strategies IB, Software verification and development cost. http://www.ibs-inc.net
69. Synopsys: synopsys system-level catalyst. http://www.synopsys.com/
70. Tasiran S, Yu Y, Batson B (2004) Linking simulation with formal verification at a higher level. IEEE Des. Test 21(6):472–482. doi:10.1109/MDT.2004.94
71. Vermeulen B, Goossens K (2011) Interactive debug of socs with multiple clocks. Des Test Comput IEEE 28(3):44–51. doi:10.1109/MDT.2011.42
72. Wehner P, Ferger M, Göhringer D, Hübner M (2013) Rapid prototyping of a portable hw/sw co-design on the virtual zynq platform using systemc. In: SoCC. IEEE, pp 296–300

73. Winterholer M (2006) Transaction-based hardware software co-verification. In: FDL'06: Proceedings of the conference on forum on specification and design languages
74. Zeller A (2005) Why programs fail: a guide to systematic debugging. Morgan Kaufmann
75. Zhao M, Liu Z, Liang Z, Zhou D (2009) An on-chip in-circuit emulation architecture for debugging an asynchronous java accelerator. In: International conference on computational intelligence and software engineering, CiSE 2009, pp 1–4. doi:10.1109/CISE.2009.5363421

Chapter 2
Embedded Software Debug in Simulation and Emulation Environments for Interface IP

Cyprian Wronka and Jan Kotas

2.1 Firmware Debug Methods Overview

Present EDA environments [1, 2] provide various methods for firmware debug. Typically one can use one of the following:

- Simulation with a SystemC model of the hardware. This allows for a very early start of firmware development without any access to hardware and allows to test the functionality of the code assuming the model is accurate. The main limitations are lack of system view and (depending on the model accuracy) lack of hardware timing accuracy (behavioral models).
- Hardware simulation with firmware executing natively on the simulator CPU. This is the simplest method incorporating the actual RTL that allows to prototype the code. It requires some SystemC wrappers to get access to registers and interrupts. It lacks the system view and therefore cannot verify the behavior of the firmware in the presence of other system elements.
- Playback (with ability to play in both directions) of a recorded system simulation session.
- Hardware simulation with a full system model. (This is a synchronous hybrid, where RTL and software are run in the same simulation process). This can be divided into:
  - Using a fast model of the CPU [3]—this allows very fast execution of code (e.g., Linux boot in ~1 min) but lacks cycle accuracy due to TLM to RTL translation. It also slows down significantly when the full RTL simulation starts (all clocks enabled). Example of such a system is presented in Fig. 2.1.
  - Using a full system RTL—this is generally very slow and only allows to test simple operations (under 10 k CPU instructions) in a reasonable time.

C. Wronka (✉)
Cadence® Design Systems, San Jose, CA, USA
e-mail: cwronka@cadence.com

J. Kotas
Cadence® Design Systems, Katowice, Poland
e-mail: jank@cadence.com

D. Lettnin and M. Winterholer (eds.), *Embedded Software Verification and Debugging*, Embedded Systems, DOI 10.1007/978-1-4614-2266-2_2

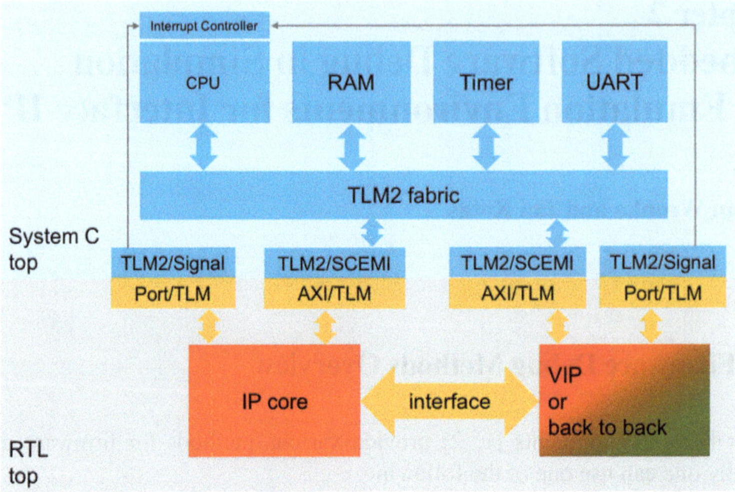

Fig. 2.1 Diagram showing a generic hybrid hardware simulation or emulation environment with a TLM part including the CPU fast model (*top*) and the RTL part including an interface IP core (*bottom*)

- Hardware emulation [4] of the full system. Again this can be divided into:
  - Hybrid mode consisting of a fast model CPU and emulation RTL. In the case of interface IP it provides very fast execution, but requires good management of memory between the fast model and the emulation device to assure that the data transfers (typically data written by CPU and later sent to interface) will be efficiently emulated. *NOTE: In this mode software is executing asynchronously to RTL and the two synchronize on cross domain transactions and on a set time interval. Effectively the software timing is not cycle accurate with the hardware and depending on setup would remove cache transactions and cache miss memory transactions.*
  - Full RTL mode where all system is cycle accurate. This is slower (Linux boot can take 10 min), however consistent performance is maintained through the emulation process. This mode allows to test the generic system use cases or replicate problems found during FPGA testing.
  - Emulation with no CPU—A PCIe SpeedBridge® Adapter can be used to connect an arbitrary interface IP device to a PC and develop a driver in the PCIe space. The emulation environment allows for access to all internal signals of the IP (captured at runtime, even using a very elaborate condition-based trigger) to debug the issues (whether originating from, software, hardware or the device connected at the other end of the interface.

- Hardware prototyping using FPGA. In this case the processor can run at 10–100 s of MHz (or even at GHz speeds if it is a silicon core connected to the FPGA logic).

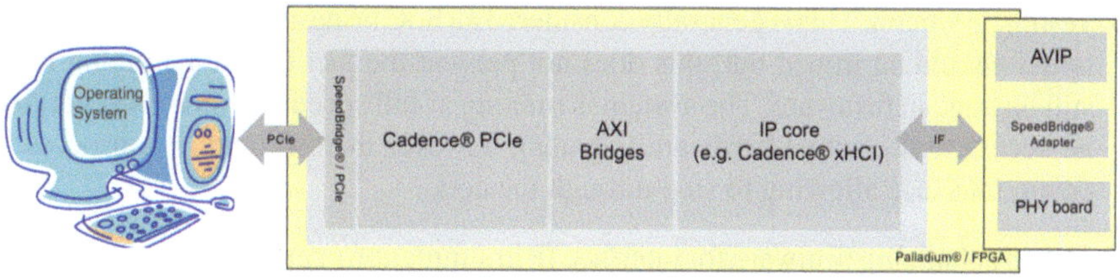

Fig. 2.2 Example setup of a PCIe device used for Linux driver debug with Cadence® Palladium® platform

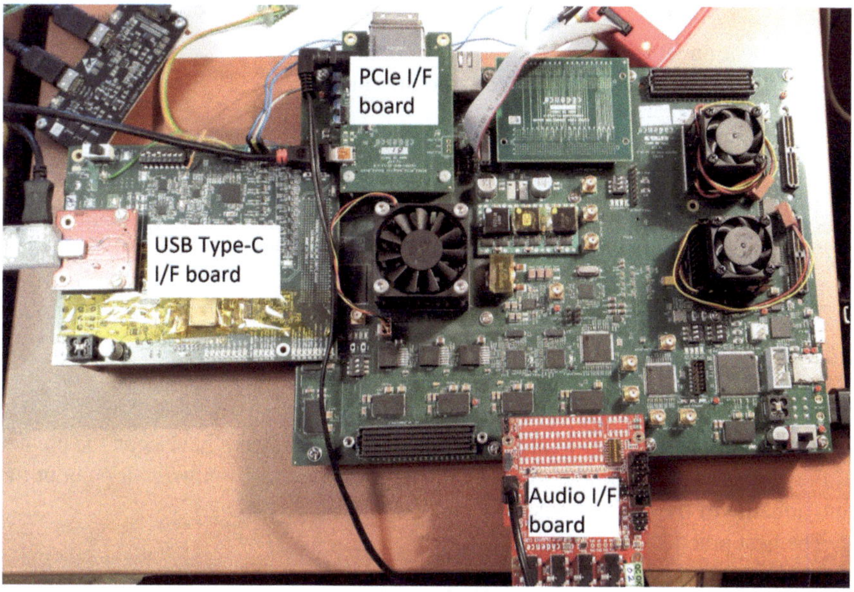

Fig. 2.3 Example Cadence® FPGA board setup used for interface IP bring-up. Please note the number of connectors and the attached boards. This is required for testing systems consisting of multiple interface IPs connected together (also with a CPU and system memory)

These environment are not very good at bringing up new (unproven) hardware, however they are great for:

- System tests in real time or near real time
- Performance tests
- Soak tests.

Example schematic used both for FPGA and Emulation or RTL interface IP connected to a standard PC is shown in Fig. 2.2. Alternatively it is possible to prototype a simple SoC in an FPGA. Such a system with PCIe EP (End Point) IP, USB Device IP, and an audio interface is presented in Fig. 2.3.

- Testing in silicon. This is generally considered an SoC bring-up, where all hardware issues should be ironed out, this does not prevent the fact that some tuning may still be done in firmware. The system is running at full speed and there is generally no access to the actual hardware (signals), however there should be a processor debug interface allowing to step through the code.

When debugging firmware for interface IP in simulation or emulation, it is required to connect the interface side of the IP to some entity that is compliant with the interface protocol. To achieve this one can use:

- In simulation:
  - Verification IP
  - Another instance of IP with the same interface
- In emulation:
  - Accelerated Verification IP
  - Another instance of IP with the same interface
  - A SpeedBridge® Adapter to connect to the real world.

2.2 Firmware Debuggability

In many cases the engineer looking after the firmware may not have been the creator of the code, and therefore it is important to provide as much information about the functionality of the code as possible.

The best practices include:

- Self-explaining register name structures or macros
- Self-explaining variable names (as opposed to a, b, c, d)
- Self-explaining function names
- Especially useful is information about the usage of pointers, as often firmware uses pointers to store routine addresses, it is important to provide sufficient information for a debugging engineer to be able to understand what the code is supposed to do.

There are new systems appearing currently on the market, where majority of IP driver code can be automatically generated from a higher level (behavioral) language [5]. In that case it is important to be able to regenerate the code and trace all changes back to the source (meta code) in order to propagate all debug

```
DESCRIPTION = ''Sequence for starting playback.'';
sequence start_playback
{
  LOCAL unsigned int idx;

  idx=0;
  repeat(idx<8)
```

```
  {
    /* If a channel has been enabled, start playback on that
     channel*/
    if (DRV_I2S_CID_CTRL.I2S_STROBE[idx] == 0)
    {
      DRV_I2S_CTRL.ENB[idx] = 1;
    }
    idx = idx+1;
  }
  /* Enable clock to transmit sync */
  DRV_I2S_CID_CTRL.STROBE_TS = 1;
  DRV_I2S_CID_CTRL.STROBE_RS = 0;
  DRV_I2S_CTRL.TFIFO_RST = 0;

  /* Reset */
  DRV_I2S_CTRL.TSYNC_RST = 1;
}
```

Listing 1 Example code using VayavyaLabs DPS language.

Generated code:

```
/*!
 * \brief Sequence for starting playback.
 * \return Success or Failure
 * \retval  0 Success
 * \retval -1 Failure
 */

INT I2S_MC_IP_SLAVE_start_playback(void)
{
  UINT idx;
  UINT varDRV_I2S_CID_CTRL;

  idx = 0;

  while (idx < 8) {
    DRV_I2S_CID_CTRL_RgRd(varDRV_I2S_CID_CTRL);
    if (GET_VALUE(varDRV_I2S_CID_CTRL,
          DRV_I2S_CID_CTRL_I2S_STROBE_LPOS,
          DRV_I2S_CID_CTRL_I2S_STROBE_HPOS) == 0)
      DRV_I2S_CTRL_ENB_UdfWr(idx, 1);
      idx++;
    }

  DRV_I2S_CID_CTRL_STROBE_TS_UdfWr(1);
  DRV_I2S_CID_CTRL_STROBE_RS_UdfWr(0);
  DRV_I2S_CTRL_TFIFO_RST_UdfWr(0);
  DRV_I2S_CTRL_TSYNC_RST_UdfWr(1);

  return Y_SUCCESS;
}
```

Listing 2 Example code generated using VayavyaLabs DDGen

Using register abstraction as described above allows to detach the code from the register and filed addresses and if any register moves in the structure, the code is not affected.

2.3 Test-Driven Firmware Development for Interface IP

Working with a new piece of IP that is usually just being developed as the initial firmware is created, requites a constant closed- loop work mode in which new versions of code can be tested against new versions of RTL. A typical development flow within an interface IP firmware support team could look like this:

- Hardware team designs the IP
- Firmware engineers participate in the design to feed on the register interface design decisions
- Once a register model is designed and a first RTL implementation is in place with a functional bus connectivity, such IP can immediately be integrated into early firmware development.

One of the interesting methodologies to use in firmware development in test-driven development. This could be considered as a formalization of the hardware bring-up process, where some expectation of functionality is always well defined, and the development/bring-up aims at getting that feature enabled/supported.

The steps of test driven development are

- Design a test that fails
- Implement functionality to pass
- Refine for ease of integration.

2.3.1 Starting Development

The first cycle of development is typically hardware focused. The initial test is generally a register read/write operation on the address space of the IP to confirm that it has been properly integrated with the system Fig. 2.6.

Once this code is in place (and since it is always the same, it is easily reusable) the preparation of the test platform can start. In current hardware simulation environments, it is possible to use a Virtual Platform Environment (VPE) type environment (diagram) where the CPU and all system peripherals exist as SystemC fast models and the IP under development is being connected through a TLM to RTL wrapper. In a typical use case such connection requires the following steps:

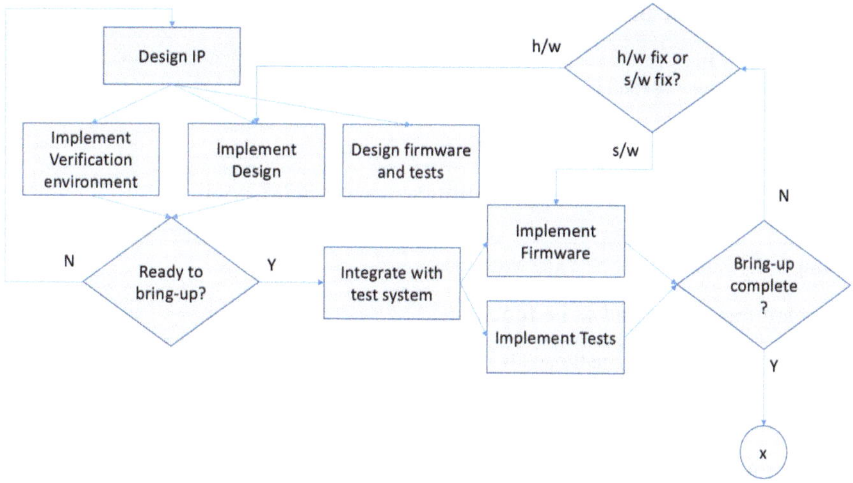

Fig. 2.4 Diagram showing cooperation between RTL (h/w) and C (s/w) teams to design and bring-up an IP block

- Preparation of a system-level wrapper for the new IP
- Integration of that wrapper with the existing system
  - Selecting base address
  - Providing any control signals that are required by that IP but not available on the available slave bus interfaces
  - In the case of interface IPs it is crucial to connect the actual 'outside world' interface of the IP (diagram) to a sensible transactor. These can be:

    An instance of Verification IP

    An instance of a IP core compatible with the interface.

In an ideal world such integration should be seamless as the IP comes with standard bus interfaces and only requires these connected to the systems. In reality the 'transactor' part of the test environment can be the most laborious element of the test environment preparation. In early stages this can be delayed until first contact with IP registers has been made, but this typically is a stage that can be passed very quickly (Fig. 2.4).

Once all is connected and a binary file is loaded into the system, a couple of software/hardware co-debug cycles encompassing:

- checking slave bus ports access,
- checking interrupt generation

lead to a first iteration of a working firmware debug environment.

Example C code of a register read and write operation:

```
log(‘‘reading:\n’’);
v=PAA_UncachedRead32((volatile uint32_t*)(IP_REGS_BASE+
    offset));

log(‘‘writing:\n’’);
PAA_UncachedWrite32((volatile uint32_t*)(IP_REGS_BASE+i), 0
    xFABEDD1E);
```

Listing 3 Example register read and write code

with bare-metal read and write functions:

```
uint32_t PAA_UncachedRead32(volatile uint32_t* address) {
    return *address;
}

void PAA_UncachedWrite32(volatile uint32_t* address,
    uint32_t value) {
    *address = value;
}
```

Listing 4 Bare-metal implementation of platform abstraction API.

An example of a register read and write operation on a slave port is presented on Fig. 2.6.

At this stage the firmware debug process typically starts to cross its paths with the hardware verification team. If the IP comes with a machine-readable register description (such as IP-XACT or SysRDL) again it is straightforward to automatically generate system-level test sequences that can be executed immediately on the new design with the benefit of the test code being fully driven by the register design data (Fig. 2.5). This constitutes a very basic set of system sanity test cases such as:

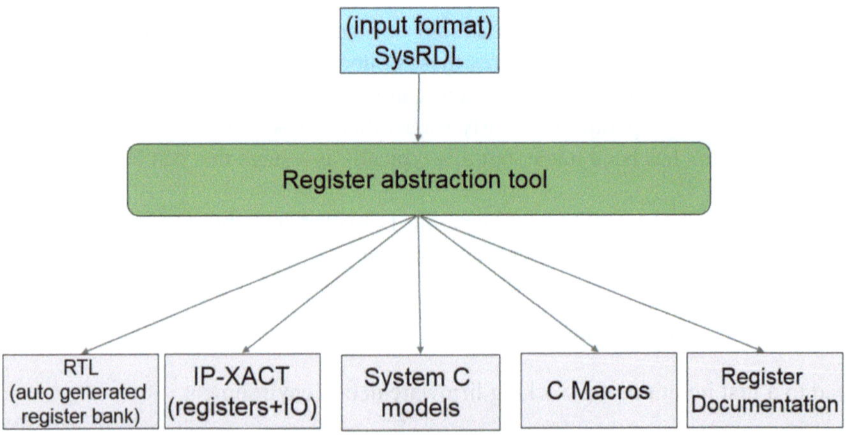

Fig. 2.5 Simplified register abstraction flow for interface IP

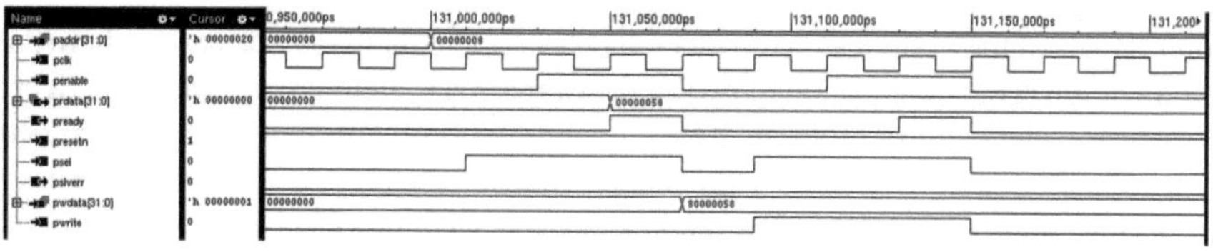

Fig. 2.6 Reading and writing register at address 8 (on APB)

- check register reset value
- confirm register writable bits are writable
- confirm register read only bits are read only.

These are generally covered by the hardware verification team, however the execution of these in a real system allows ironing out issues with bus access and any minor register bank problems.

At this stage a firmware developer can start running any functional test cases on the IP core.

2.3.2 *First Functional Tests*

In a typical situation of developing a low-level driver for a certain IP there is a number of 'key features' that need to be supported to allow for initial system-level integration. These typically are a subset of:

- Initialize IP,
- send data over interface using register interface,
- receive data over interface using register interface,
- use the master bus interface/trigger a DMA,

2.3.2.1 Initializing IP

Once a verification environment is in place, one can set up a simple routine to perform the initialization of the IP and confirm that it has been performed correctly. A test to confirm that may be

- receiving an interrupt that system is ready
- polling a register that system is ready
- in extreme cases it may require to peek into the actual IP signals to see that it has been intialized.

IP initialization may involve setting up elements connected to the IP to establish a working data channel, e.g., PHY initialization. At that stage it is important for all components of the system to be connected properly and there may be cases where the initialization procedure executed by the firmware in a system environment uncovers a hardware connectivity problem. These integration problems can be easily avoided by using automated IP assembly tools [6]. In these cases it is also required to have a working 'transactor' at the other end of the 'external interface' to assure that link has been established. If the transactor is another instance of IP supporting the same protocol, it is desirable to have a unit test environment, where each IP can be tested in isolation against a known model (e.g., a Verification IP instance).

2.3.2.2 Sending and Receiving Data

Once the IP is initialized and a data channel is in place, it is possible to start transmitting data to confirm the system connectivity. In the case of interface IP it is required to have a 'transactor' instance able to fulfill the communication protocol requirements to assure that the IP can complete the data transmission. Let's consider an example of a flash memory controller. The aim of the debug step is to confirm that we can create a working code able to write a page of data in the NAND memory by sending the data word by word through the register interface. To achieve this it is important to have the following prerequisites:

- the controller is initialized
- the memory is initialized
- it is required to know the protocol used by the memory controller (in the case of NAND memory it is a sequence of standardized commands (e.g., ONFI) interleaved with the desired data.

Let us consider an example operation where a memory page is written using register interface and later read to confirm consistency. Debugging this code in simulation, allows the following:

- stepping through the firmware code
- looking at the register interface to confirm the right data is sent to the flash controller
- looking at the memory interface to confirm that the right data is sent and received to/from the flash memory (model)
- looking at the register interface to confirm that the right data is read from the memory (model)
- accessing a memory (or having debug messages) to confirm the data integrity.

Once the code has been debugged and proven in simulation, it can be tested on an FPGA setup. Once all basic issues are resolved in simulation, the FPGA platform allows for performance analysis and profiling of the opeartions using real memory parts.

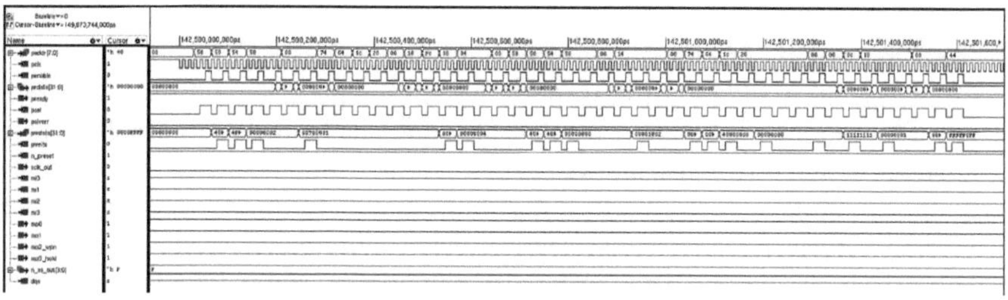

Fig. 2.7 Performing an initialization sequence on the APB bus

2.3.2.3 Testing the DMA

The simulation environment is a great platform to test the DMA transfers and to prototype the firmware to perform these. DMA transfers can show system problems (e.g., with memory caching) and therefore it is good to be able to see the system memory at the time of the transfer being initiated and the flow of data on the master bus interface to confirm the integrity of the data being transmitted. The diagram below shows the same memory write and read as in previous section, but this time a descriptor requesting each operation is built and after the initial trigger, the IP is expected to:

- read the descriptor
- start reading the memory (using a master bus interface)
- start writing the data to the device
- finish reading memory
- finish writing the data
- interrupt the processor to notify the operation has completed
- perform steps similar to above, but in the other direction.

Once a single DMA has been confirmed to work, it may be required to test a chained DMA operation, in which multiple descriptors are fetched (in a sequence, potentially with partial hardware buffering) to assure the firmware is capable of creating and handling these (Fig. 2.7).

Another route may be to test a multi-channel DMA where multiple chains of commands are created and executed by the interface IP on multiple connected devices.

2.3.2.4 Sending and Receiving Data on FPGA

In the case of an FPGA the debug capabilities are significantly reduced, the number of IP signals that can be traced is limited and having checked the code in simulation should significantly reduce the bring-up process. Having an embedded processor in the FPGA connected to the debugger still allows to step through the code, however looking at the interface signals can be difficult and may require the use of an oscilloscope or a logical analyser.

The earlier defined steps (already proven in simulation) can be repeated on FPGA to bring-up the basic functionality of the hardware.

The advantage of FPGA testing is speed and for example in the case of NAND flash testing it was possible to run over 2 orders of magnitude more data (500 MB) than in simulation (16 MB). This also allowed to detect issues with the driver with handling large number of data blocks. Further, due to the fact that testing was performed on real memory it was possible to discover that the memory model used was erased by default, which was not the case for the real memory die, and required an update in the memory erase procedure (Fig. 2.8).

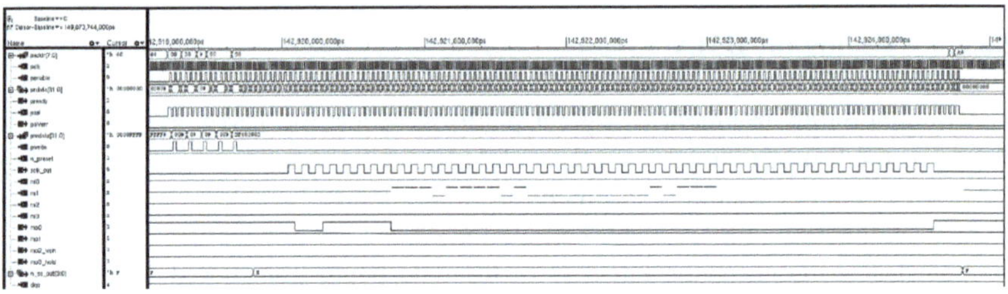

Fig. 2.8 Enabling the interface on a QSPI IP

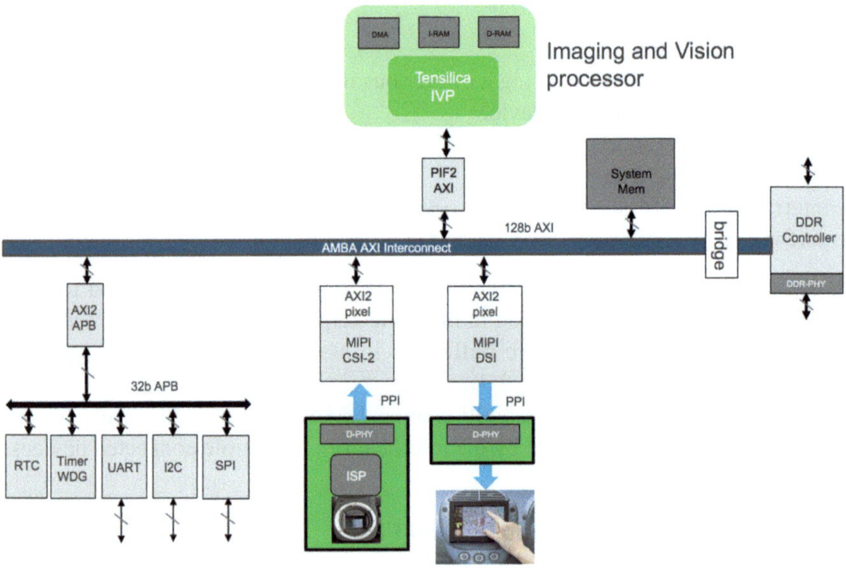

Fig. 2.9 Example system with an Image Processing Unit and a camera/display MIPI interface

2.3.3 Debugging a System

Once all components of the system have been brought up (at least in the positive path) it is possible to start debugging the system along with applications that use multiple components at the same time. The assumptions at this point should be:

- there is a debuggable CPU (allowing to step through code)
- all components can perform basic operations
- the CPU and all opponents have access to memory.

Once these steps are established one can proceed to system bring-up. In most cases such bring-up would start with some basic bare-metal tests, where multiple component would be used at the same time. For example the UART may need to be used as a debug output (or even input) while bringing up a camera interface (MIPI CSI) IP. Additionally it may be required to initialize the camera using another simple interface (e.g., I2C) while using the MIPI interface to obtain images and place them into the system memory. In that case it is important to understand the full system architecture Fig. 2.9 as multiple components will be interacting with each other and the CPU will be interacting with all components.

The debug steps in the example above could follow as:

- create a serial port connection for debug messages (run UART)
- create an I2C connection with camera and log connection results
- initialize camera and log the results
- initialize the MIPI CSI interface
- request the MIPI CSI interface to start transferring data from the camera to the system memory.
- access the memory using a debugger or dump its contents using the UART.

In this example, it would be desirable to have a controlled environment in which the camera would send known data that can be verified (e.g., test mode), whether the correct content has been written into the system memory. Once the system is confirmed to perform all operations in the right sequence, and the right data is copied into memory, it is also required to confirm that the data transfer is well synchronized with the rest of the system, i.e., that once the camera starts sending multiple frames per second, there is sufficient buffering and the frame data is not corrupted. This is a good opportunity to test the interrupts as multiple devices (UART, I2C, MIPI CSI) can trigger an interrupt and it is important to set up the right way of processing interrupts. Example of prioritized interrupt execution is shown in Fig. 2.10.

Interrupt handling should not be affecting the functionality of the system and therefore it may be required to create artificial conditions that trigger multiple interrupts at the same time to debug the scenario (Fig. 2.11).

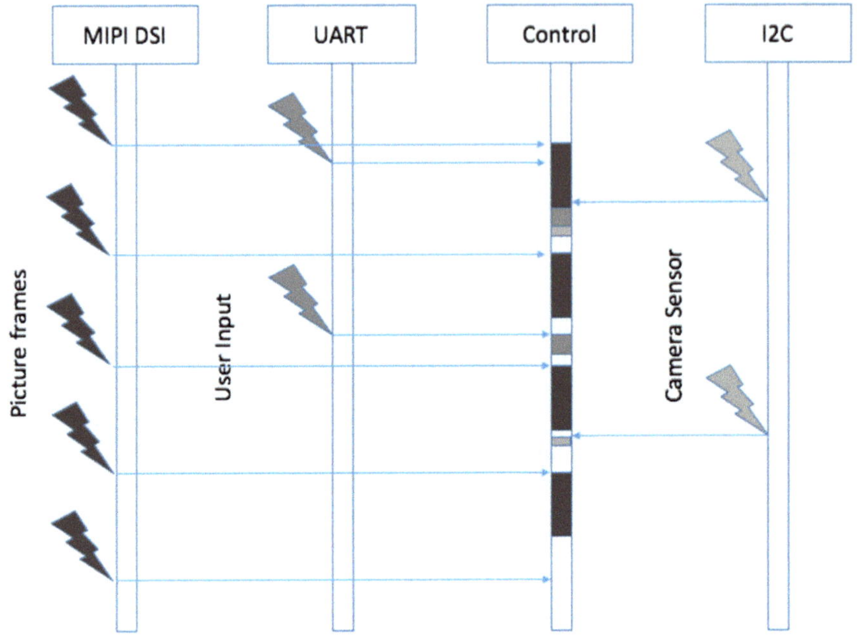

Fig. 2.10 Diagram showing multiple devices generating interrupts at various times. The interrupt priority needs to be set appropriately to assure no interruption in streaming data and user interaction can be compromised in this case, as it is only for set-up and debug purposes

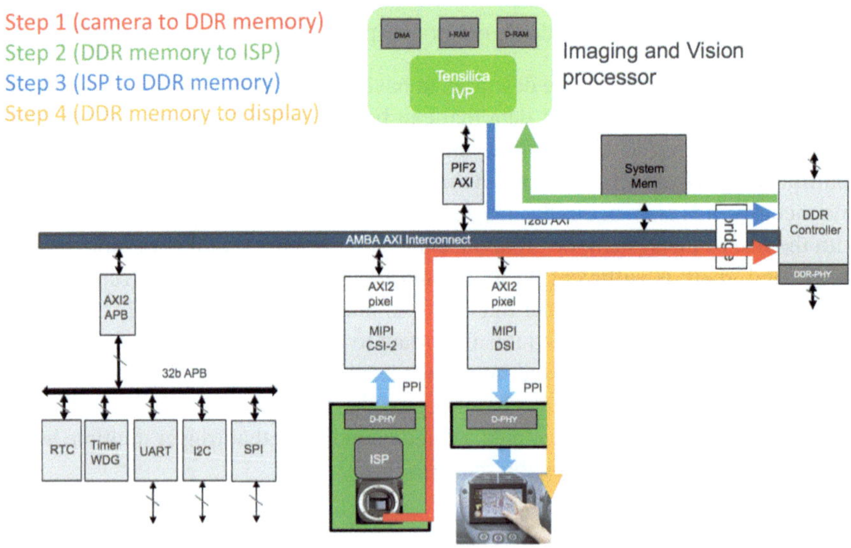

Fig. 2.11 Diagram showing memory transfers (DMA) in the system

2.3.4 System Performance

Once all components are working together it is possible to consider the system performance and look at code optimisations to achieve some desired benchmarks. Let's consider the system above with the addition of a screen interface and an image processing routine. In that case the image is

- fetched form the camera and stored into memory,
- read from that memory by an Image/Signal processing component,
- stored back to memory by the Image/Signal processing component
- read from memory by the display component (e.g., MIPI DSI).

Considering a VGA test image is processed in RGB mode the system will need to read 2 MB and write 2 MB of data per image. If that image is delivered at 30 fps, this is 60 MB/s read and 60 MB/s written. One has to consider that the system may need to access the memory at the same time for other purposes (e.g., logging/maintenance/reading program or other data related to image processing. If the system is running on an FPGA it is required to assure the system architecture enables the required bandwidth. And the FPGA platform itself is an ideal vehicle to prove that this can be achieved.

Once the system is performance proven in FPGA, the scalability to ASIC speeds should be much more simple and only minor software optimizations may be required. The issues to be considered here would be:

- bus access
- memory access
- memory bandwidth (bus, controller, module)
- memory caching.

Lab example

"In one of our demo designs, the team was presented with an existing FPGA board where the main (and biggest) FPGA chip did not have a direct connection to the DDR memory. The DDR controller was synthesized on a separate (smaller FPGA) that could be connected to the main one using a chip2chip connection. The system described above (with camera, image processor and display) was implemented on that board and the memory bandwidth turned out to be a limiting factor in terms of image resolution and frames per second. A special design using a Xilinx Chip2 AXI construct was used to achieve best performance. This needed further optimizations both on hardware and software level. On the hardware side the important tuning factors were:

- chip2chip clock speed (and synchronous/asynchronous implementation versus the system bus)
- image processor cache size
- image processor cache line size (tuned with the DDR line size).

On the software side it was important to limit memory access and assure that the image blocks (in each on the four data paths) would be sent using a 'chained' sequence of DMA operations to best utilize the fabric bandwidth."

2.3.5 *Interface IP Performance in a Full Featured OS Case*

Another interesting example could be the Ethernet MAC implementation used as a NIC (network interface card) with a PCIe endpoint. Both cores are connected together and plugged into a PC (diagram below). There are two mirroring set-ups with:

- emulation (using a PCIe SpeedBridge® Adapter and Ethernet SpeedBridge® Adapter)
- FPGA with standard PCIe and Ethernet connectors.

Either of these implementations can be connected to a lab PC running Linux OS and a driver implemented for the interface IP as a PCIe device in Linux can be performance tested and debugged. (The same set-up is used for compliance testing).

In a particular case a feature was added to the Ethernet controller to allow offloading the software TCP/IP stack from performing 'coalescing' of received packets into one large packet. Support for this feature was added to the Linux driver and both implementations of the NIC platform were used to test the functionality in a system environment. The FPGA implementation was used to benchmark the CPU savings due to the offload feature and while running these a potential problem with the offloads was discovered. The following steps were used to track the problem with FPGA):

- Additional logging in the Linux driver
- Enabling additional kernel logging in Linux
- Using a TCP sniffer (wireshark) to confirm data transfer.

However all these methods did not allow to find the cause of the problem, and therefore the problem needed to be replicated in an emulation environment for full understanding. Moreover, it turned out that some of the problems are inconsistent from one PC to another. After replicating the problem in an emulation environment with full access to all hardware design signals it turned out that:

- The PCIe root-port (host) may reorder packets
- Some PCIe bus parameters influence the occurrence of this issue
- Further to all these, an actual hardware issue was detected with further refinement of the test, that was not related to the PCIe bus.

In the case of performing such lab tests, it is very important to have a very strict control over the test environemnt, to assure no random events or changes in the set-up cause the system to behave in a different way. The best practice is to set up a clean sandbos with a fully scripted path through the tests and a full control over external events (e.g., avoid connecting to live local networks).

2.3.6 *Low Level Firmware Debug in a State-of-the-Art Embedded System*

While bringing up a new SoC system in the lab used for compliance testing purposes, the hardware designers created a system consisting of a 64-bit CPU core and a contemporary interrupt controller, both used to test the Message Signal Interrupt (MSI) capability of an embedded xHCI controller 2.1 with xHCI controller as IP and a USB SuperSpeed device VIP instance. The MSI capability is typical and well supported for PCIe-based devices, however with the introduction of new interrupt controllers it is possible to receive these interrupts in embedded systems.

In the case of an embedded MSI interrupt the memory location is hijacked by the interrupt controller to generate an interrupt signal for the CPU. This has required to set the correct configuration of the bus (including protected access line) to work with the interrupt controller and allow devices on the bus to access the given memory location to generate an interrupt. An example device supporting this feature is an xHCI interface core. The Linux driver supports the MSI mode and with the full system running in the simulation environment it was possible to confirm that the IP generates MSI interrupts. After appropriate configuration of addresses, switching the xHCI core into the MSI mode and configuring the GIC to receive these messages.

2.4 Firmware Bring-up as a Hardware Verification Tool

With the current possibility to bring-up an IP quickly in a simulation/emulation in a full system environment, it is possible to use off-the-shelf system tests for early verification of newly developed IP.

This has been applied to two different cores recently in our lab setting.

2.4.1 *NAND Flash*

An MTD (Memory Technology Device) driver was created for a new NAND flash controller core to enable the usage of MTD-test framework in Linux to perform multiple (existing) system test on an IP that was still in development.

List of MTD tests available in the community[1]

- **mtd_speedtest**: measures and reports read/write/erase speed of the MTD device.
- **mtd_stresstest**: performs random read/write/erase operations and validates the MTD device I/O capabilities.

[1] http://www.linux-mtd.infradead.org/doc/general.html.

- **mtd_readtest**: this tests reads whole MTD device, one NAND page at a time including OOB (or 512 bytes at a time in case of flashes like NOR) and checks that reading works properly.
- **mtd_pagetest**: relevant only for NAND flashes, tests NAND page writing and reading in different sizes and order; this test was originally developed for testing the OneNAND driver, so it might be a little OneNAND-oriented, but must work on any NAND flash.
- **mtd_oobtest**: relevant only for NAND flashes, tests that the OOB area I/O works properly by writing data to different offsets and verifying it.
- **mtd_subpagetest**: relevant only for NAND flashes, tests I/O.
- **mtd_torturetest**: this test is designed to wear out flash eraseblocks. It repeatedly writes and erases the same group of eraseblocks until an I/O error happens, so be careful! The test supports a number of options (see `modinfo mtd_torturetest`) which allow you to set the amount of eraseblocks to torture and how the torturing is done. You may limit the amount of torturing cycles using the `cycles_count` module parameter. It may be very good idea to run this test for some time and validate your flash driver and HW, providing you have a spare device. For example, we caught a rather rare and nasty DMA issues on an OMAP2 board with OneNAND flash, just by running this tests for few hours.
- **mtd_nandecctest**: a simple test that checks correctness of the built-in software ECC for 256 and 512-byte buffers; this test is not driver-specific, but tests general NAND support code.

This allowed finding a number of issues with the Linux driver, as it was a ready set of 'tested tests' so a perfect environment for driver development with a well defined acceptance criteria (passing all tests).

The MTD framework allows to run file-systems (such as JFFS2) on top of the existing stack, giving further system and stress testing capabilities.

2.4.2 xHCI

In the case of an IP with a standard register interface, the Linux itself can be used as a first stage of system testing as soon as the IP can be integrated into a simulation/emulation environment. Furthermore, the USB-test Linux framework was used to perform multiple system tests to confirm the system-level stability of an IP (that was at the same time going through design updates and standard hardware verification process).

```
TEST 1:  write length bytes iteration times
TEST 2:  read length bytes iteration times
TEST 3:  write/length i*vary..length bytes iteration times
TEST 4:  read/length i*vary..length bytes iteration times
Queued bulk I/O tests
TEST 5:  write iteration sglists sglen entries of length
    bytes
```

```
TEST 6:  read iteration sglists sglen entries of length
    bytes
TEST 7:  write/vary iteration sglists sglen entries i*vary..
    length bytes
TEST 8:  read/vary iteration sglists sglen entries i*vary..
    length bytes
Non-queued sanity tests for control (chapter 9 subset)
TEST 9:  ch9 (subset) control tests, iteration times
Queued control messaging
TEST 10:  queue sglen control calls, iteration times
Simple non-queued unlinks (ring with one urb)
TEST 11:  unlink iteration reads of length
TEST 12:  unlink iteration writes of length
EP halt tests
TEST 13:  set/clear iteration halts
Control write tests
TEST 14:  iteration ep0out, 0..length vary vary
Iso write tests
TEST 15:  write iteration iso, sglen entries of length bytes
Iso read tests
TEST 16:  read iteration iso, sglen entries of length bytes
Tests for bulk I/O using DMA mapping by core and odd address
TEST 17: write odd addr length bytes iterations times core
    map
TEST 18: read odd addr length bytes iterations times core
    map
Tests for bulk I/O using premapped coherent buffer and odd
    address
TEST 19: write odd addr length bytes iterations times
    premapped
TEST 20: read odd addr length bytes iterations times
    premapped
Control write tests with unaligned buffer
TEST 21: iterations ep0out add addr, 1|0..length vary vary
Unaligned iso tests
TEST 22: write iterations iso odd, sglen entries of length
    bytes
TEST 23: read iterations iso odd, sglen entries of length
    bytes
Unlink URBs from a bulk-OUT queue
TEST 24: unlink from iteration queues of sglen length-byte
    writes
Simple non-queued interrupt I/O tests
TEST 25: write length bytes iterations times
TEST 26: read length bytes iterations times
Simple Bulk performance test
TEST 27: bulk write iterations*sglen*length /(1024*1204)
    Mbytes
TEST 28: bulk read iterations*sglen*length /(1024*1204)
    Mbytes
```

Listing 5 Non-queued bulk I/O tests from http://www.linux-usb.org/usbtest/

where:

- iterations—count iterations
- length—size of packet
- sglen—scather/gather entries of
- vary—vary packet size by.
- i—current iteration

These tests allowed to find a number of issues with the driver, in particular:

- stall support
- device configuration.

The test suite was also used as a stress test for the IP using several system configurations (32/64 bit, legacy/MSI interrupt) allowing to assure system quality of the driver.

2.5 Playback Debugging with Cadence® Indago™ Embedded Software Debugger

Digital systems get more and more complex in time. The requirements for new functionality while preserving backwards compatibility makes each new system release more complicated than the previous one. To fully verify such a project, a tool is required that allows firmware and hardware co-verification. Hardware tests (whether UVM or functional/directed) may not suffice, in order to create more realistic use cases firmware driven test cases are required. This allows building a complete use case into a single test case, e.g., adding a new sound channel to a SoundWire device requires reconfiguring the IP and running all transmissions simultaneously. Therefore it is important to have a system with a microcontroller available as part of the verification platform.

On the other hand, increasing competition on the market enforces putting more consideration to the cost of designing new versions of devices. In particular the time-to-market perspective is very important. It is key to assure that the work is not held up by unexpected errors that are difficult to analyze. The occurrence of such issues can lead to project delays and introduction of 'crunch time'—both significantly affecting the team and potentially leading to further problems. There is a lot of cases where the error is observable after it occurs and it is easy to set a breakpoint at the later execution time to see its effects, but it may be difficult to track the root cause (such as—unexpected input value or a bug in the code). In this case it is very important to be able to observe the past of the system (hardware and software including memory contents) to trace the actual root cause occurrence.

Indago™ Embedded Software Debugger is a solution allowing to debug the entire system (both hardware and software). It is a complete solution in the way that using a single tool allows to access all components of the project. It allows easy analysis

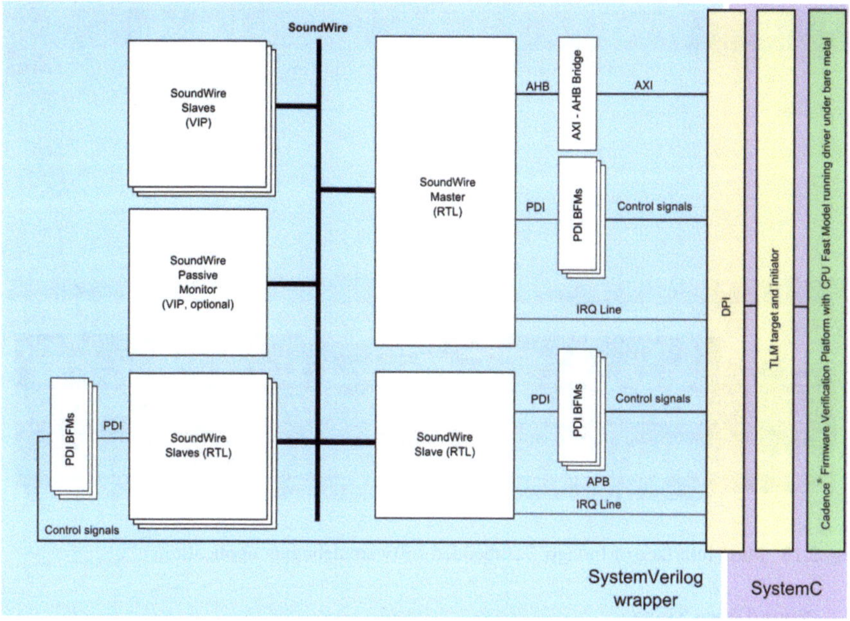

Fig. 2.12 Soundwire hardware/software co-verification platform

of the cooperation between hardware and software. Indago™ Embedded Software Debugger allows to watch the complete system state at each moment by tracing all signals, the CPU state, and the memory contents.

2.5.1 *Example*

An example verification system for the SoundWire Master and Slave devices has been created and is shown in Fig. 2.12. It is based on an SoC with both SystemC and RTL components executed in the Incisive® Enterprise Simulator.

The system consists of the following components:

- SystemC:
  - CPU Fast Model
  - Virtual Fabric Model
  - Memory Model
  - UART Model
- RTL (Verilog):
  - SoundWire Master

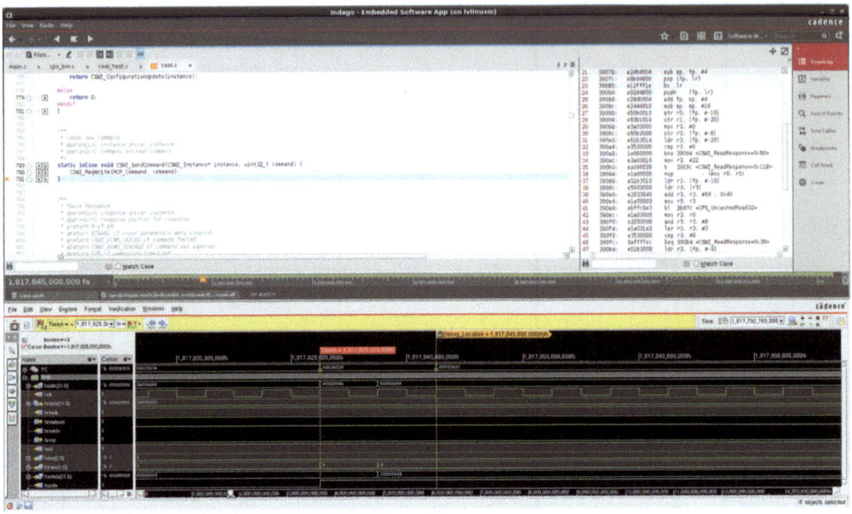

Fig. 2.13 Main Interface of Indago™ embedded software debugger application

  – SoundWire Slave

- VIP:

  – SoundWire Slave

In order to use playback debugging with Indago™ Embedded Software Debugger, the ELF file and information about software execution flow are required. It can be either:

- Trace text file which allows to use all features of Indago™ Embedded Software Debugger,
- PC Counter trace on a waveform, which only provides basic debugging functionality.

A compatible waveform database containing information about hardware signals can also be loaded to the application, which allows hardware/software co-debugging.

Figure 2.13 shows the main interface of the Indago™ Embedded Software Debugger which contains two major parts, code debugger, and waveform window which can show information about both software and hardware.

This particular example shows the write operation to the hardware. CPU writes data to the FIFO which is shown at the hardware level as AHB write transaction.

The offline debugging feature allows to move forward and backward in code execution and setting unlimited breakpoints. This is especially useful when code fails in an interrupt handler. Indago™ Embedded Software Debugger allows to jump back to the last execution and see what went wrong without stepping through successful runs. Figure 2.14 presents an example function which was called by an interrupt handler. Call Stack and waveform can be used to quickly identify why it was called.

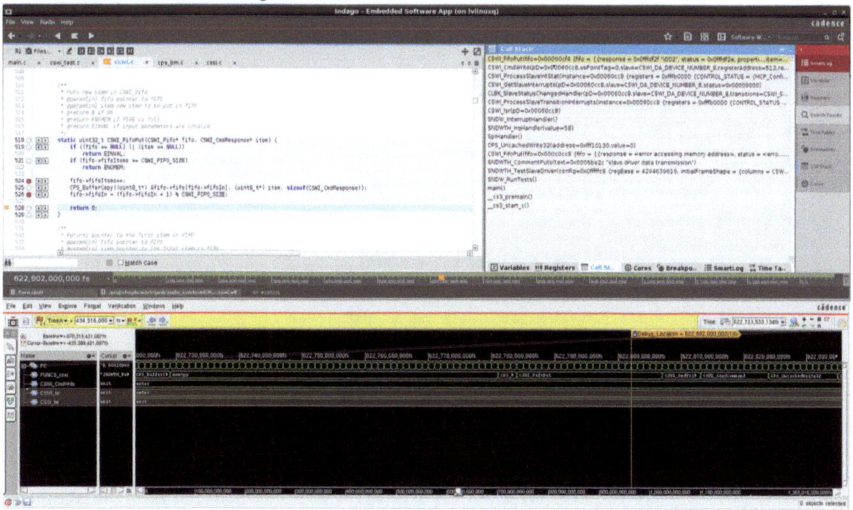

Fig. 2.14 Debugging of code executed during one of interrupts

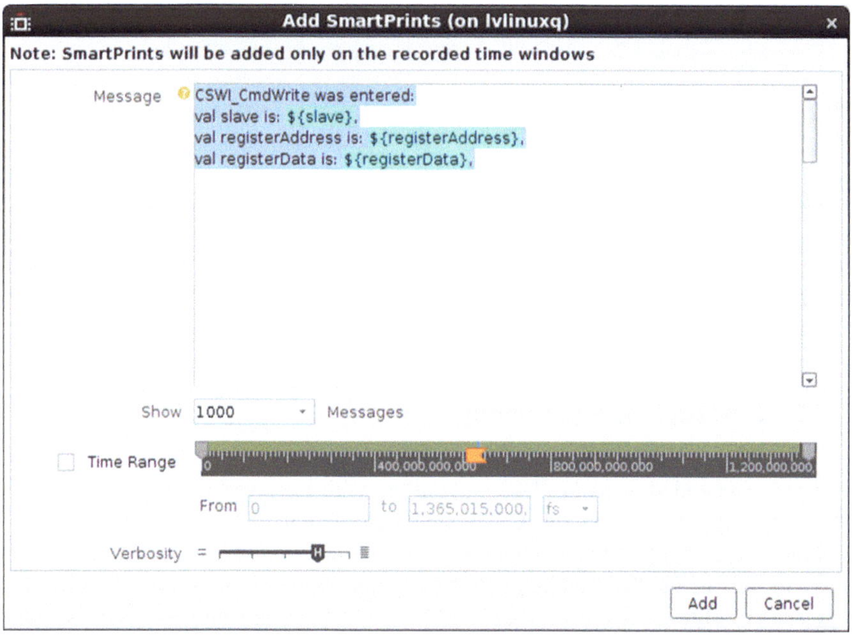

Fig. 2.15 SmartPrint settings

Offline debugging allows nonintrusive debugging methods. Using prints to debug code is quite common, but it affects code execution timing, so problematic code may start working after adding some prints in the code. Indago™ Embedded Software Debugger supports SmartPrints. In any line of code which was executed it is possible

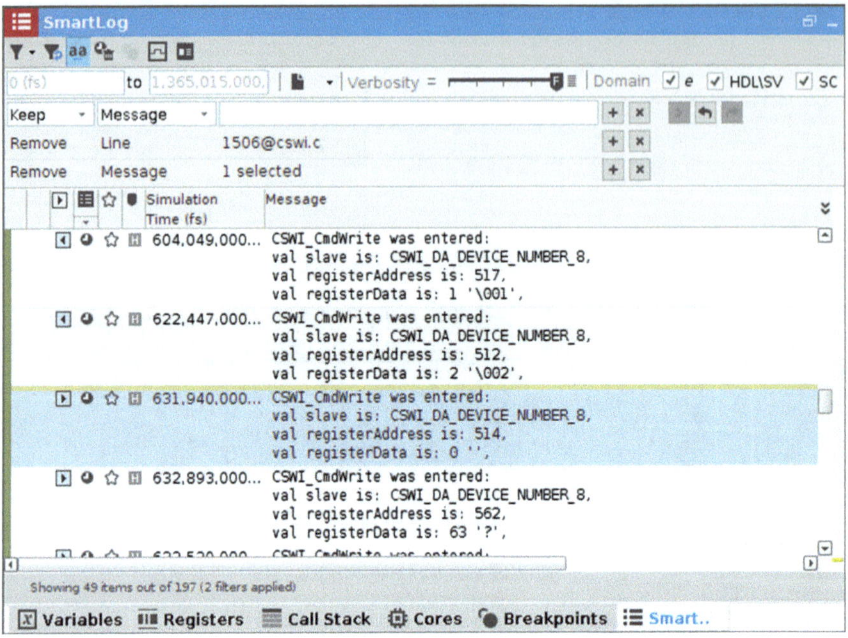

Fig. 2.16 SmartLog window

to add print message with custom information and available variables, as shown in Figs. 2.15 and 2.16. Because it uses offline database of software execution, it does not add any delays, which may affect the execution especially during interrupts. It also makes debugging much easier.

2.5.2 Coverage Measurement

Indago™ Embedded Software Debugger along with Incisive® Enterprise Simulator Metrics Center allows for Code Coverage measurement.

Figure 2.17 shows the main window of the application. It provides a quick insight into code testing quality, and helps to identify and fix issues in the code verification. It is also possible to view code source code with indication which line was executed as shown in Fig. 2.18. Results from multiple test runs can be merged together.

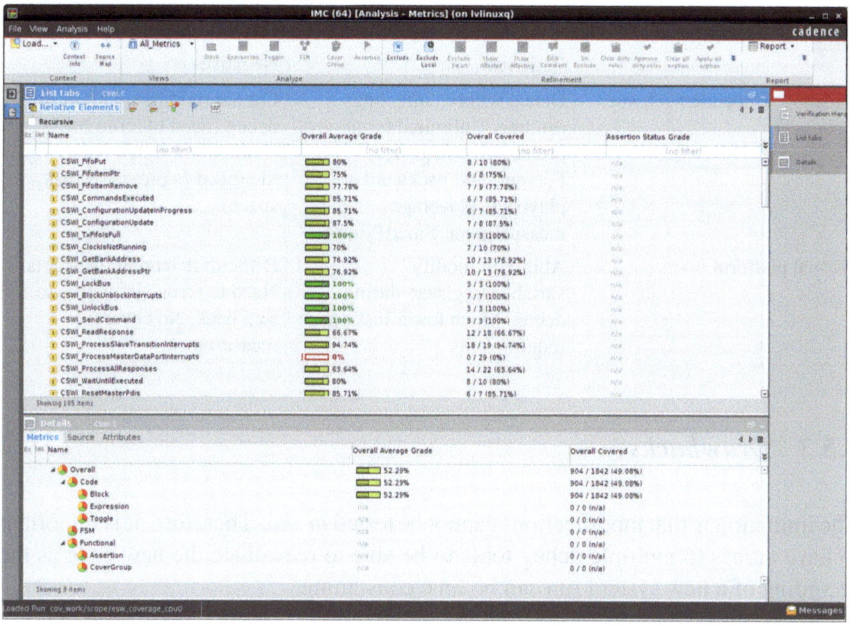

Fig. 2.17 Incisive® Enterprise simulator metrics center presenting example code coverage results

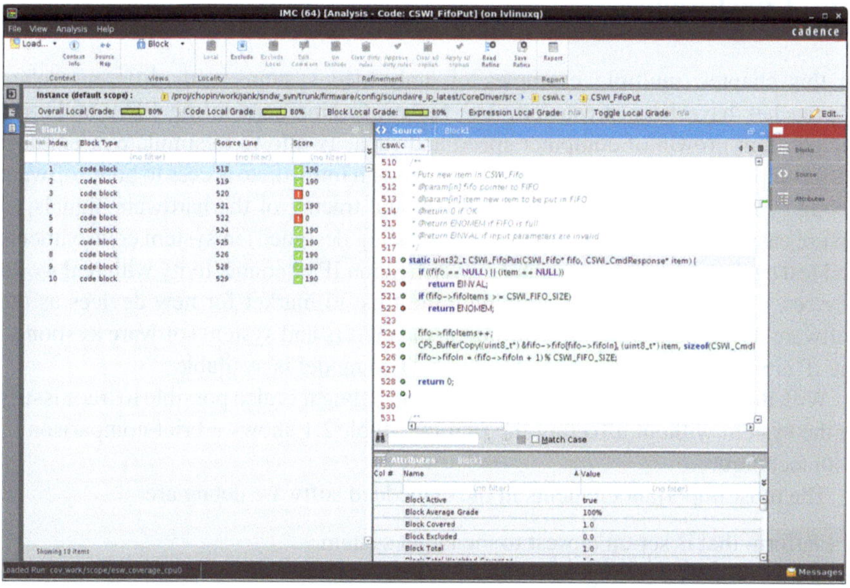

Fig. 2.18 Example function source coverage view

Table 2.1 Comparison of Cadence® runtime and playback debuggers

	Positives	Negatives
Indago™ embedded software debugger	Debug without affecting runtime. Unlimited rerun of prerecorded sequence. Forward and backward playback. Coverage measurement. SmartPrints	Cannot modify values during debug (need to rerun full simulation). Large recording file (need to provision disk space)
Virtual platform	Ability to modify variables/registers during debug. Much lower disk space requirements	Difficult debuging interrupts. Need to rerun simulation to step back. No coverage measurement

2.5.3 Drawbacks

The limitation is that modifications cannot be tested *in situ*. Therefore, it is important to have access to multiple debug tools to be able to reevaluate the new code as the recording of a new system run can be time consuming.

2.6 Conclusions

In this chapter, multiple examples of embedded systems with different debug approaches have been presented. The recent development in the EDA tools and the constant growth of computer speed and capacity allows to simulate or emulate very complex system on chip- type platforms with immediate access to both software debug symbols and stepping through code and tracing of the hardware signals, to assure easy bring-up of new devices (while being designed) in system environments and test against both industry standard verification IP or connectivity with real world devices. This in turn allows shortening the time to market for new devices as the software developers can start working on the drivers and system software as soon as the IP is *reasonably* stable or as soon as a TLM model is available.

With the new **playback** approach to system debug it is also possible to trace issues in the system without affecting the runtime. Table 2.1 shows a brief comparison of both techniques.

The most important elements in the embedded software debug are

- Platform that is set-up closest to the target system.
- Ability ot start working with the device or its model as early as possible in the development process.
- Cooperation of hardware and software engineers (especially in the early bring-up stages).
- The appropriate choice of tools.

Acknowledgements The authors would like to thank Rafał Ciepiela for his input into the contents and review of the chapter.

References

1. Cadence® (2016) Cadence Incisive® Enterprise simulator. Product website. https://www.cadence.com/content/cadence-www/global/en_US/home/tools/system-design-and-verification/simulation-and-testbench-verification/incisive-enterprise-simulator.html
2. Synopsys (2016) Synopsys virtualizer. Product website. http://www.synopsys.com/Prototyping/VirtualPrototyping/Pages/virtualizer.aspx
3. ARM (2016) ARM cpu fast models. Product website. http://www.arm.com/products/tools/models/fast-models.php
4. Cadence® (2016) Cadence Palladium® Z1 emulator. Product website. https://www.cadence.com/content/cadence-www/global/en_US/home/tools/system-design-and-verification/acceleration-and-emulation/palladium-z1.html
5. Vayavya Labs (2016) Vayavya device driver generator. Product website. http://vayavyalabs.com/technology/ddgen/
6. ARM (2016) ARM socrates DE tool. Product website. https://www.arm.com/products/system-ip/ip-tooling/socrates-design-environment.php

Chapter 3
The Use of Dynamic Temporal Assertions for Debugging

Ziad A. Al-Sharif, Clinton L. Jeffery and Mahmoud H. Said

3.1 Introduction

The growth in the software industry is rapid and the size of programs is becoming larger and larger. In contrast, the rate of advances in the debugging literature is relatively slow. Most debuggers are well suited for a specific class or set of bugs. Program bugs can be caused by numerous circumstances and revealed long after their root cause. Understanding the source code and the execution behavior of the program is essential to locate and find the cause of most bugs. This understanding can be achieved by different means; one is to employ different debugging sessions that capture, depict, analyze, and investigate the state of the program at, and in between, different points of execution.

A typical interactive source-level debugger is one of the most valuable debugging tools, but it relies heavily on the user's ability to conduct a live investigation. It helps programmers locate and find the root cause of bugs by stepping through the source code and examining the current state of execution.

Source-level debugging techniques such as conditional breakpoints and watchpoints are dynamically inserted during the debugging session. They can check execution properties and stop the execution whenever a condition is satisfied. Even though such breakpoints may have the advantage of being conditional and dynamic with on-the-fly *insertion*, *deletion*, and *modification*, they are still bounded to their locations; the exact line number in the source code of the target program and the current state of the referenced variables and objects at that location on that execution time. For instance, a class variable may be assigned a bad value in a method that is not on the stack when the bug that caused the crash or core dump is revealed. This may force

Z.A. Al-Sharif (✉) · M.H. Said
Software Engineering Department, Jordan University of Science and Technology, P.O. Box 3030, Irbid 22110, Jordan
e-mail: zasharif@just.edu.jo

M.H. Said
e-mail: mhsaid@just.edu.jo

C.L. Jeffery
Computer Science Department, University of Idaho, Moscow, ID 83844, USA
e-mail: jeffery@uidaho.edu

D. Lettnin and M. Winterholer (eds.), *Embedded Software Verification and Debugging*, Embedded Systems, DOI 10.1007/978-1-4614-2266-2_3

a user to run multiple debugging sessions on the same bug before it is understood. Typically, a user can investigate the current state. If there is no evidence of the bug's root cause, he/she may restart the execution hoping to stop at an earlier point where the cause of the bug is still accessible [6]. In contrast, Temporal Assertions are logical expressions that use Temporal Logic (TL) in order to validate, not one state, but a sequence of execution states, such as a sequence of variable values changed within a block of code [8–10].

In order to introduce Dynamic Temporal Assertions (DTA) into conventional source-level debugging sessions, for this research a source-level debugger named UDB [1, 4] was extended with on-the-fly temporal assertions (made from within the live debugging session). UDB is the source level debugger for the Unicon programming language; it is packaged with the Unicon language distribution on Source Forge and downloaded from *unicon.org*. Aside from the temporal assertions extension, UDB's command set is that of GDB. UDB was used instead of GDB for this research because its higher level execution monitoring abstractions allow UDB to be more easily extended than is GDB [2, 3].

The new DTA assertions that UDB supports are not bounded by the limitations of ordinary breakpoints such as *locality* and *temporality*. UDB's DTA assertions serve three purposes:

1. Extend the usability of conventional source-level debuggers' conditional breakpoints and watchpoints. This simplifies the ability to validate relationships that may extend over the entire execution and check information beyond the state of evaluation.
2. Reduce the number of times a user has to stop and single step the execution for state-based investigation.
3. Augment a traditional breakpoint-based debugging session with testing and verification capabilities [7].

3.1.1 DTA Assertions Versus Ordinary Assertions

Standard in-code assertions are inserted into the source code to validate pre- and post-conditions or to check the value of some variables and expressions. In general, typical ordinary assertions suffer from three limitations:

- *Locality*: An ordinary in-code assertion is bounded by its location (scope); it cannot reference a variable from another scope even if it is live based on the current execution state. Assertions live in one of the functions; each can reference local and global variables. If the scope is a method, it can reference any of the class variables. In fact, typical assertions cannot check or validate local variables in other functions or methods, even if that foreign local is static or still live somewhere on the stack of the current program's execution state. For example, what if a user needs to check the value of variable `x` from `procedure foo()` against variable `y` from `procedure bar()`? see Figs. 3.1 and 3.2.

```
procedure foo( )
    local a, b, c
    x := 10
    ..............
    bar()
    ..............
end
procedure bar( )
    local a, b, c
    y := 20
    ...............
    y := ( y * a ) / b – c
    // virtually assert always() { y >= foo:x }
    ...............
end
```

Fig. 3.1 The possibility of a temporal assertion over two live procedures

```
procedure foo( )
    static x := 0
    x +:= 1
    ..............
end
procedure bar( )
    static y := 0
     y +:= 1
    ...............
end
procedure baz( )
    foo()
    ...............
    bar()
    // virtually assert always() { bar:y >= foo:x }
    ...............
end
```

Fig. 3.2 The possibility of a temporal assertion over two sibling procedures

- *Temporality*: An ordinary in-code assertion is bounded by the current state of execution. It can check only the current value of the referenced variables. For example, what if a user needs to check the value of variable x against the previous or even the initial value of x? Ordinary assertions are found to be useless once more.

- *Dynamicity*: An ordinary in-code assertion is bounded to the source code, where it is written and compiled; any change or modification requires the ability to recompile and rebuild the executable. If the ordinary assertion evaluates to `false`, it may provide a warning statement or terminate the execution. If the user wants to investigate, he/she may modify the assertion by tightening or loosening the condition, or adding nearby assertions.

3.1.2 DTA Assertions Versus Conditional Breakpoints

Conditional breakpoints and watchpoints are dynamically inserted during the debugging session. They can check execution properties and stop the execution whenever a condition is satisfied. Even though such breakpoints may have the advantage of being conditional and dynamic with on-the-fly *insertion*, *deletion*, and *modification*, they are still bounded to their locations; the exact line number in the source code of the target program and the current state of the referenced variables and objects at that location on that execution time. Whereas, DTA assertions are able to reference variables that are not accessible (not active in the current execution state) at evaluation time. This feature solves the problem provided in Fig. 3.2, which shows that procedure `foo()` and `bar()` are siblings in `baz()`.

3.2 Debugging with DTA Assertions

In general, users of source-level debuggers suffer from:

1. The limited information provided about the execution history, and
2. The lack of automated trace-based and analysis-based debugging techniques, which may help users validate various execution states.

DTA assertions, within a typical source-level debugger, provide an extension of conditional breakpoints and watchpoints. They employ agents that implement temporal logic operators, each with an automatic tracing mechanism. Traced data are *assertion-driven*; relevant information is gathered and analyzed in real time. Different DTA assertions can be applied on different execution properties with dynamicity and flexibility. Each assertion is capable of validating program properties that may extend over a sequence of execution states [14].

For example, a debugging process may include checking variable values from different scopes. Figure 3.3 shows a program that prints out the prime numbers from `1` to some `x`. The `procedure main()` calls `isPrime()`, which returns `true` when the passed argument is a primary number. The temporal assertion provided in #1 of Fig. 3.3 shows how to check the current local value of variable `i` against the last value of variable `i` of the `procedure main()` (denoted by `main:i`). This DTA assertion assumes that the value of parameter `i` should not change during the

```
procedure main()
    local x, i

    writes(" Please enter a positive integer number : ")
    x := read()

    write("\n The following are the primary numbers <= ", x)
    every i := 1 to x do
       if isPrime( i ) then
          write( i , " is a primary number ")
end

procedure isPrime( i )
    local k

    k  := i
    i  -:= 1
                                        assert always() { i == main:i }   #1
    while ( i > 1 ) do
    {
       if  k % i  = 0  then
          fail
       i -:= 1
    }
    return k
end
```

Fig. 3.3 Using temporal assertions to check variables from various scopes

execution of `isPrime()`. However, because the program is modifying the value of `i`, this assertion will evaluate to `false` at every change to `i` (temporal-state) in this `isPrime()` function, and it will evaluate to `false` at every return from this `isPrime()` function (temporal-interval).

3.3 Design

DTA assertions do not replace traditional breakpoints or watchpoints, instead they provide a technique to reduce their number, which means they are used to reduce the number of execution stops and improve the overall process of investigation. These temporal assertions advance breakpoints with agents of temporal logic operators (temporal agents). At a stop, besides the source-level debugging functionalities, the user can *delete*, *enable*, *disable*, and *modify* existing assertions, or even *insert* new

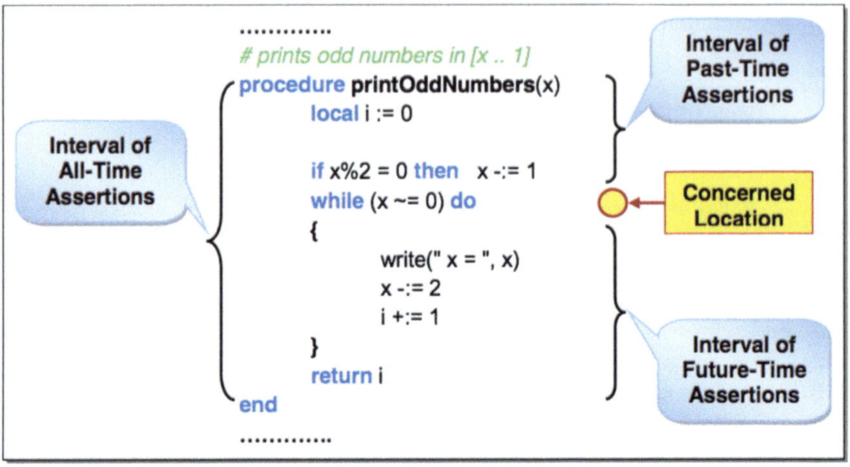

Fig. 3.4 Temporal assertions: scope and interval

Table 3.1 Atomic data related agents

Agent name	Return type	Description
initial(*x*)	Any	The initial value of *x*
final(*x*)	Any	The final value of *x*
old(*x*)	Any	The previous value of *x*
current(*x*)	Any	The current value of *x*
new(*x*)	Any	The next value of *x*
max(*x*)	Numeric	The maximum of all *x* values
min(*x*)	Numeric	The minimum of all *x* values
newmax(*x*)	True/false	Evaluate *True* if *x* has new max, *False* otherwise
newmin(*x*)	True/false	Evaluate *True* if *x* has new min, *False* otherwise
sum(*x*)	Numeric	The sum of all *x* values
avg(*x*)	Numeric	The average of all *x* values

DTA assertions at any location in the buggy program source code; all without the need to recompile the target program source code or to reload it under the debugger.

UDB supports three kinds of DTA assertions, see Fig. 3.4. Each of these kinds has its own set of temporal agents. All these DTAs can reference execution properties and other internal extension agents such as the atomic data agents described in Table 3.1 and the behavioral agents described in Table 3.2.

Table 3.2 Atomic execution behavior-related agents

Agent name	Return type	Description
call(proc)	Integer	The number of times *proc* is been called
return(proc)	Variable	The current value returned by *proc*
initialized(x)	True/false	*True* if *x* was assigned at first reference, *False* otherwise
dead(x)	True/false	*True* if *x* is never referenced at least once, *False* otherwise
reference(x)	Integer	The number of times *x* is been read and written
assign(x)	Integer	The number of times *x* is been assigned
read(x)	Integer	The number of times *x* is been read only
alias(x)	List	All current *x* aliases
iterations(loop)	Integer	The number of actual iterations of *loop*

3.3.1 *Past-Time DTA Assertions*

This category consists of four Past-Time Operators. These operators utilize information retained between entering an assertion's scope and a reaching assertion's source code location. At insertion time, the debugger starts retaining relevant information to be used during the assertion's evaluation. When the execution reaches the virtual execution point, where the assertion is hooked in the buggy program space, the assertion temporal interval is evaluated. If the evaluation is not able to complete due to some missing information (maybe out-of-scope referenced data is never used during an assertion's lifetime), the assertion evaluation is tagged with `Not Valid`. These four DTA assertions are:

1. `alwaysp() {expr}`: asserts that an expression must always hold (evaluate to true) for each, temporal state, temporal interval, and during the whole execution.
2. `sometimep() {expr}`: asserts that an expression must hold at least once for each temporal interval, and during the whole execution.
3. `previous() {expr}`: asserts that an expression must hold right at the last state before the end of the temporal interval.
4. `since() {condition ==> expr}`: asserts that an expression must hold right after condition is true up until the end of the temporal interval and for each interval.

3.3.2 *Future-Time DTA Assertions*

This category consists of four Future-Time Operators. These operators utilize information retained between reaching an assertion's source code location and leaving an assertion's scope. The agents of those operators start watching for referenced objects when the evaluation is triggered, where the debugger starts retaining relevant information until the assertion's temporal interval is evaluated completely. If the execution is terminated before the assertion's interval is complete, the user is able to check temporal states in that incomplete temporal interval. These four DTA assertions are:

1. `alwaysf() {expr}`: asserts that an expression must always hold (evaluate to true) for each, state, temporal interval, and during the whole execution.
2. `sometimef() {expr}`: asserts that an expression must hold at least once for each temporal interval, and during the whole execution.
3. `next() {expr}`: asserts that an expression must hold right at the very first state in the temporal interval.
4. `until() {condition ==> expr}`: asserts that an expression must hold from the beginning of the temporal interval up until `condition` is true or the end of the temporal interval and for each interval.

3.3.3 All-Time DTA Assertions

This category consists of two All-Time Operators. These two operators are based on the time interval between entering an assertion's scope and exiting an assertion's scope. When the assertion scope is entered, the assertion starts retaining relevant information and evaluates its temporal states. When the execution exits the assertion scope, the assertion temporal interval is evaluated. These two DTA assertions are:

1. `always() {expr}`: asserts that an expression must always hold (evaluate to true) for each, state, temporal interval, and during the whole execution.
2. `sometime() {expr}`: asserts that an expression must hold at least once for each temporal interval, and during the whole execution.

3.4 Assertion's Evaluation

Each reached (evaluated) assertion has at least one temporal interval. This interval consists of a sequence of temporal states. Temporal interval is defined by the assertion scope and kind. Assertion's scope is defined based on the source code location provided in the `assert` command. This scope is the procedure or method surrounding the assertion location. Figure 3.4 shows the temporal interval for all three kinds of temporal assertions in reference to the provided location. Together, the assertion's scope and kind define the temporal interval. In particular:

- *Temporal Intervals of Past-Time* DTA assertions start at entering the assertion's scope (calling the scope procedure) and end at reaching assertion's source code location for the very first time after entering the scope.

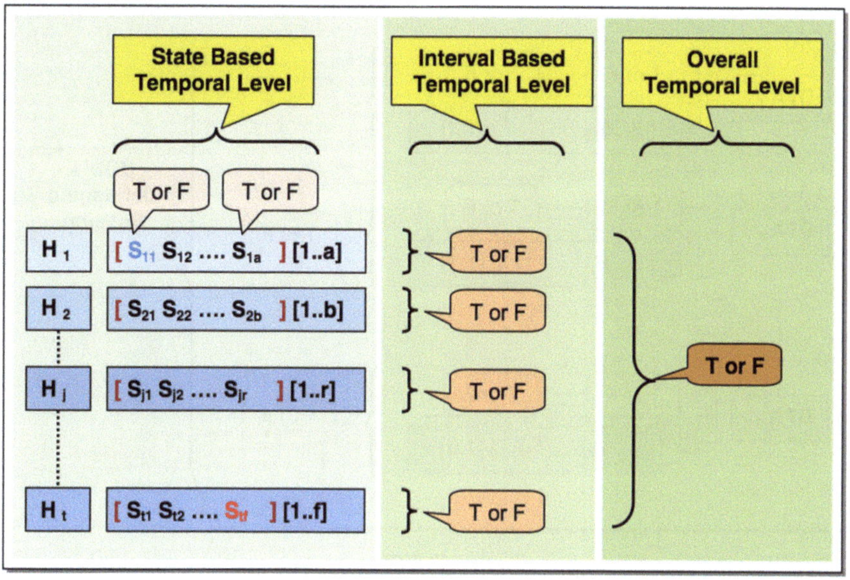

Fig. 3.5 Sample Temporal Assertion's Evaluation: An assertion is hit t times $[H_1..H_t]$. Each hit represents a Temporal Interval, which consists of a various number of states; each state is evaluated to `True` or `False`. Each Temporal Interval is evaluated based on the conjunctive normal form of its state-based evaluations (on that particular hit H_i). Finally, on the overall temporal level, the assertion is evaluated once more based on the conjunctive normal form of all previous interval-based evaluations

- *Temporal Intervals of Future-Time* DTA assertions start at reaching an assertion's source code location for the very first time after entering the assertion's scope and end at exiting the assertion's scope (returning from the scope procedure). In this kind of temporal assertions, the source code location can be hit more than once before the interval is closed.
- *Temporal Intervals of All-Time* DTA assertions start at entering assertion's scope and end at exiting that scope; regardless of the provided source code location.

During a debugging session, it is possible for a user to have multiple assertions, each with multiple temporal intervals, and each interval with multiple temporal states. See Figs. 3.5 and 3.6. Each DTA assertion runs through three levels of evaluations:

1. *State-based*: temporal level (single state change). This evaluation is triggered by any change to the assertion referenced objects.
2. *Interval-based*: a sequence of consecutive states. This evaluation is triggered by reaching the end of assertion's temporal interval (exiting the assertion scope).
3. *Overall execution-based*: a sequence of consecutive temporal intervals. This evaluation is triggered by the end of execution.

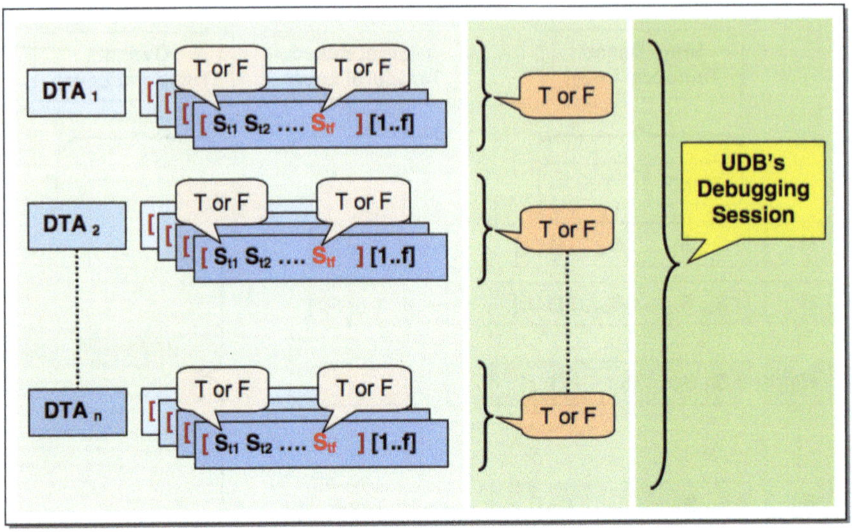

Fig. 3.6 Sample evaluation of various temporal assertions (`n DTA` assertions) during a debugging session

UDB's DTA assertions are evaluated in the debugger side. By default, whenever an assertion evaluates to `false`, the source-level debugger stops execution in a manner similar to a breakpoint. The debugger transfers control to the user with an evaluation summary.

3.4.1 Temporal Cycles and Limits

A temporal `cycle` defines the maximum number of consecutive temporal intervals (maximum number of temporal level evaluation times), which defines the overall evaluation. The default value of cycle is unlimited number of evaluations. Temporal `limit` defines the maximum number of temporal states considered in each temporal interval. The definition of temporal limit is changed based on the kind of temporal assertion in reference. In particular:

1. In Past-Time DTA assertions: `limit` defines the maximum number of consecutive states before reaching assertion's source code location and after entering the assertion's scope.
2. In Future-Time DTA assertions: `limit` defines the maximum number of consecutive states after assertion's source code location is reached and before exiting the assertion's scope.

3. In All-Time DTA assertions: `limit` defines the maximum number of states before and after an assertion's source code location is reached, all within the assertion's scope.

The default limit is defined by whatever temporal states (temporal-interval) are encountered during the execution of an assertions' scope and based on its temporal interval. The user can reduce the number of temporal states considered in each temporal interval by setting this limit using the `limit` command.

3.4.2 Evaluation Log

Furthermore, the assertion's log gives the user the ability to review the evaluation behavior of each assertion (evaluation history). The debugger maintains a hash table for each assertion. It maps assertions' intervals into lists with information about their temporal state base evaluation. Each list reflects a temporal interval, which maintains the evaluation order and result for each temporal state. Each list reflects one temporal interval, also maintained based on their order. Completely evaluated intervals are tagged with `True` or `False`. If the evaluation process is already started, but the final result is still incomplete, perhaps the end of the interval is not reached yet, these intervals are tagged with `Pending` until they are complete. This will convert `Pending` into `True` or `False`. However, some assertions may never be triggered for evaluation; this may occur because the execution never reached the assertion's insertion point during a particular run. These assertions have the hit counter set to zero.

3.4.3 DTA Assertions and Atomic Agents

Atomic agents are a special kind of extension agents (nontemporal logic agents) [1, 3, 4]. They expand the usability of DTA assertions and facilitate the ability to validate more specific data and behavioral relationships over different execution states, see Tables 3.1 and 3.2. When an atomic agent is used within a DTA assertion, it retains and processes data and observes behaviors in relevance to the used assertions. The assertion scope is what determines when the agent should start to work and what range of data it should be able to retain and process. For example, if the assertion uses the `max(var)` or `min(var)` atomic agents, the agent always retains the maximum or minimum respectively over the assertion temporal interval.

Those atomic agents add more advancement and flexibility to the usefulness of DTA assertions and their basic temporal logic operators. In particular, DTA assertions that reference atomic agents can easily check and compare data obtained by these atomic agents, which encapsulate simple data processing such as finding the minimum, maximum, sum, number of changes, or average. For example, suppose

```
1. (udb) assert test.icn:50 sometimep() { x < y }

2. (udb) assert test.icn:50 alwaysp() { old(x) != current(x) }

3. (udb) assert test.icn:50 alwaysf() { return(foo) > 0 }

4. (udb) assert test.icn:50 always() { iteration(while) < 100 }

5. (udb) assert test.icn:50 always() { call(baz) < 1000 }
```

Fig. 3.7 Sample of different UDB's temporal assertions

that a static variable is changed based on a conditional statement where it is incremented when the condition is `true` and decremented when the condition fails. What if the user is interested in the point at which this variable reaches a new maximum or minimum? DTA assertions provide a simple solution for such situations.

For example, the assertion number 1 of Fig. 3.7 will pause the execution when variable `x` becomes greater than or equal to `y`. As another example, suppose the user is interested in the reasons behind an infinite recursion; perhaps a key parameter in a recursive function is not changing. DTA assertions provide a mechanism to retain the parameter value from the last call and compare it with the value of the current call, see assertion number 2 of Fig. 3.7. If `old(x) == current(x)`, the assertion will stop the execution and hand control to the debugger where the user can perform further investigation. Of course, there are other reasons that may cause infinite recursion, such as the key parameter value changing in the opposite directions on successive calls.

Moreover, DTA assertions simplify the process of inserting assertions on program properties such as functions' return values, and loops' number of iterations. For example, a user may insert a breakpoint inside a function in order to investigate its return value, or place an in-code assertion on the value of the returned expression. A DTA assertion provides a simpler mechanism; see assertion number 3 of Fig. 3.7. Assertion number 4 of Fig. 3.7 states that the while loop at line 50 in file `test.icn` always iterates less than 100 times. Finally, assertion number 5 of Fig. 3.7 shows how to place a DTA assertion on the number of calls to a function; the assertion will stop execution at call number 1000. This particular assertion is difficult to accomplish using conventional source-level debugging features such as breakpoints and watchpoints.

3.5 Implementation

DTA assertions are virtually inserted into the buggy program source code on-the-fly during the source-level debugging session. UDB's static information is used to assist the user and check the syntax and the semantic of the inserted assertion. Each assertion is associated with two sets of information (1) *event-based* and (2) *state-based*. The debugger automatically analyzes each assertion at insertion time in order to determine each set. It finds the kind of agents that are required to be encountered in the evaluation process. If any extension agent is used, the debugger establishes an instance of that agent and associates it with its relevant object.

The host debugger maintains a hash table that maps each assertion source code location into its related object (agent). The assertion object is responsible for maintaining and evaluating its assertion. It contains information such as (1) the parsed assertion, (2) a list of all referenced variables (3) a list with all temporal intervals and their temporal states, and (4) the assertion event mask: a set of event codes to be monitored for each assertion; this event mask includes the event masks for any of the referenced agents. Execution events are acquired and analyzed in real time. Some events are used to control the execution whereas others are used to obtain information in support of the state-based technique [2, 5].

Each assertion has its own event and value masks, which are constructed automatically based on the assertion, see Fig. 3.8. A union set of all enabled assertion event masks is unified with the debugging core event mask. The result is a set of events requested by the debugging core during the execution of the buggy program. This set is recalculated whenever an assertion is added, deleted, enabled, or disabled. On-the-fly, UDB's debugging core starts asking the buggy program about this new set of events. A change on any assertion event mask alters the set of events forwarded by the

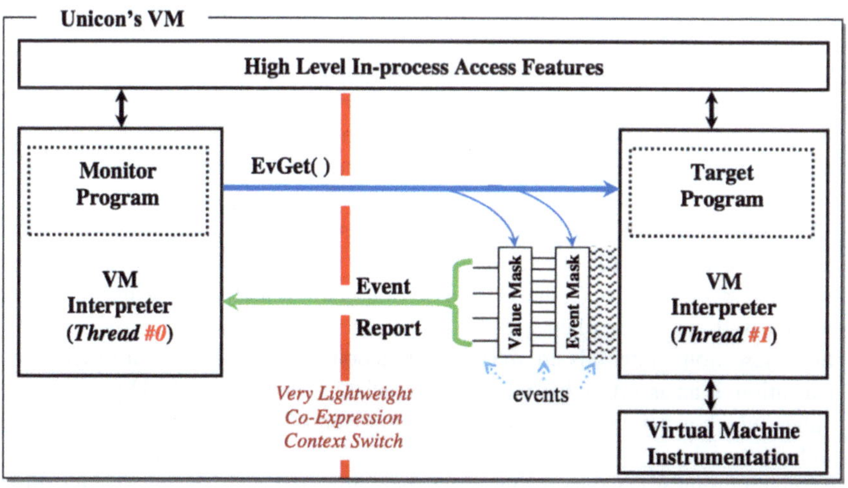

Fig. 3.8 UDB's use of event mask and event value within the Unicon virtual machine

debugging core to that assertion object. Temporal logic agents automatically obtain the buggy program state-based information to evaluate DTA assertions. Each agent automatically watches assertion referenced variables and retains their information in the debugger space.

3.6 Evaluation

DTA assertions provide the ability to validate relationships that may extend over the entire execution and check information beyond the current state of evaluation. DTA assertions' temporal logic operators are internal agents. Those agents can reference other atomic agents, which provide access to valuable execution data and behavior information. UDB's DTA assertions have the following features:

- Dynamic insertion, deletion, enabling, disabling, and modification. Assertions are managed on-the-fly during the debugging session without source or executable code alteration.
- A nondestructive way of programming supported by an assertion-free source code. In general, debugging information is needed only during program development, testing, verification, validation, and debugging.
- Assertions are virtually inserted and evaluated as part of the buggy program source code. All assertions live in the debugging session configuration; each is evaluated by the debugger in the debugger execution space. The debugger automatically maintains state-based techniques to determine what information is needed to evaluate each assertion, and it uses event-based techniques to determine when and where to trigger each assertion evaluation process. Some program state-based information is collected before assertion evaluation, while other information is obtained during the evaluation process. All DTA assertions are evaluated as if they were part of the target program space
- Optional evaluation suite, where a user can specify an evaluation action such as `stop`, `show`, and `hide`. The `show` action enriches assertions with the sense of in-code tracing and debugging with print statements, where a user can ensure that the evaluation has reached some points and the referenced variables satisfy the condition.
- The ability to log the assertion's evaluation result. This lets the user review the assertion evaluation history for a specific run. Evaluated assertions are marked with `True` or `False`. Some DTA assertions may reference data in the future; those assertions are marked with `Not Valid` for that exact state-based evaluation. Assertions' intervals are marked with a counter that tracks their order in the execution. If an assertion has never been reached, it is distinguished by its counter value, which is zero in this case. Log comparison of different runs is considered in future works.
- Most importantly, DTA assertions can go beyond the scope of the inserted location. Each assertion may refer to variables or objects that were living in the past during

previous states, but not at evaluation point, and each assertion may compare previous variable values against current or future values. Each DTA assertion implicitly employs various agents to trace referenced objects and retains their relevant state information in order to be used at evaluation time.

3.6.1 Performance

In consideration of the performance in terms of time, the implementation of temporal assertions utilizes a conservative assertion-based event-driven tracing technique. It only monitors relevant events; the event mask and value mask are generated automatically for each assertion at insertion time. Temporal assertions are evaluated in three levels. First is the state-based level, which depends on any change to the referenced execution property. Second is the interval-based level, which is determined by the assertion scope and kind. Third is the overall evaluation level, which occurs once per each execution. Different assertions can reference different execution properties. For this reason various assertions will differ in their cost.

However, in order to generally assess the role of the three evaluation levels in the complexity of these temporal assertions, let us assume that E_s is the maximum cost of monitoring and evaluating a state change within a temporal assertion. Furthermore, let us assume that n is the maximum number of state changes during a temporal interval and m is the maximum number of temporal intervals during an execution. See Figs. 3.5 and 3.6. This means, the maximum cost of evaluating a temporal interval for this assertion is $E_S{*}n$ and the maximum cost of an assertion during the whole execution is $(E_S{*}n) * m$ which is equal to $E_S{*}n * m$. However, E_s includes the cost of event forwarding. This means that part of E_s is $(2E_p + 2E_c)$, where E_c is the cost of reporting an event to UDB and E_p is the cost of forwarding an event to the temporal logic agent (internal agent). This means the E_s dominates both n and m; state change is the main performance issue in temporal assertions.

Furthermore, retained information is limited and driven by assertions' referenced execution properties. Assertions are virtually evaluated because they are in another execution space. The evaluation occurs in the debugger space with data collected and obtained from the buggy program space. The assertion log gives the user the ability to review the evaluation behavior of each assertion. Temporal assertions use in-memory tracing. A table is allocated for all assertions; it maps each assertion source code location to the instance object of the actual assertion. Another table is allocated for each assertion; it tracks temporal intervals, each of which is a list (stack) with each of the state-based evaluation result. A third table is used to map assertion temporal intervals with their evaluation result, each of which is one value `True`, `False`, or `Not Valid`. Then one variable is holding the up to the point result which is either true or false. The dominating part in the used space is the number of state changes, E_S. Each state base evaluation is tracked with a record that keeps information about the line number, file name, and the result.

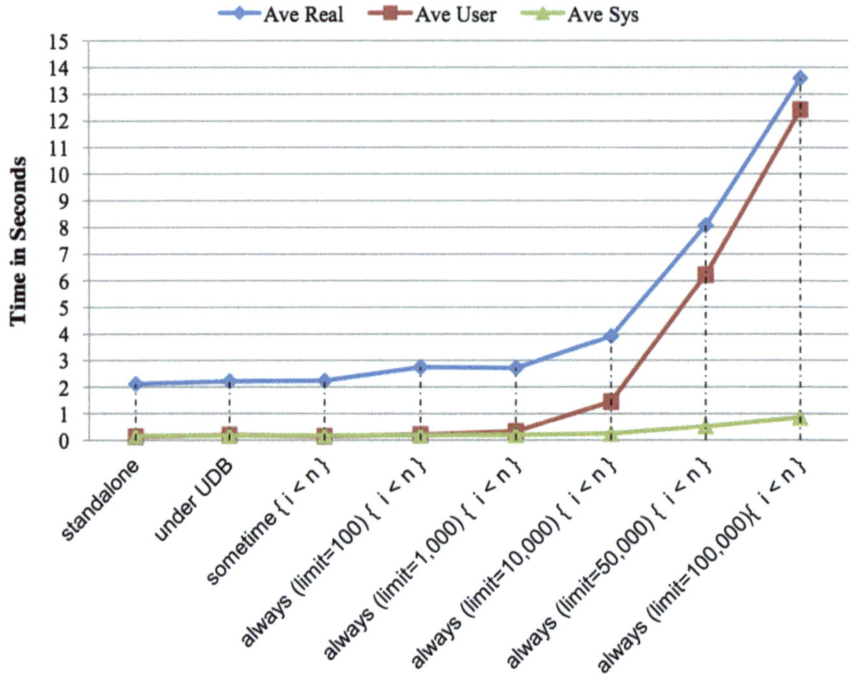

Fig. 3.9 UDB's temporal assertions evaluation time

In order to find the impact of temporal assertions on the execution of the target program and the debugging time, a simple temporal assertion is applied on a simple program. The program prints numbers between 1 and 100,000; see Fig. 3.9. The temporal assertion is applied with various sizes of temporal intervals. These intervals start at size 1, 100, 1000, 10,000, 50,000, and 100,000. The experiment is based on eight kinds of runs, each is observed for five times and the average of these times is reported. These kinds of runs range from measuring the time for the program in the standalone mode (no monitoring is involved), monitored under UDB with no assertion applied, then with an assertion that has various intervals. Figure 3.9 shows the impact of these temporal assertions on the execution time.

3.7 Challenges and Future Work

Debugging with DTA assertions provides advantages over typical assertions and conditional breakpoints and watchpoints. At the same time, it faces some challenges and limitations, some of which are based on associating assertions with the executable's source code, evaluating assertions in the debugger, and the source-level debugger's ability to obtain and retain relevant event-based and state-based information with reasonable performance.

First, if an assertion makes a reference to a variable, which is not accessible from within the assertion's scope, the debugger should automatically trace those variables and retain their relevant state information to be used at the assertion evaluation time. This allows a DTA assertion to access data that is not live at the assertion's evaluation time.

Second, what if the assertion source code location is overlapping with a statement? Which one should be evaluated first, the assertion or the statement? A conservative approach may consider the assertion evaluation after the statement only if the statement has no variables referenced by the assertion, or if the statement does not assign to any of the assertion referenced variables. However, if the statement will assign to any of the assertion referenced variable, the assertion can be evaluated before and after the statement evaluation. If the two evaluations are different such as one is `true` and the other is `false`, or both are `false`, the assertion will stop the execution and hand the control to the debugger and the user to investigate. The work presented in this paper, takes the simplest approach which is to evaluate the assertion before the statement. Furthermore, if an assertion is not overlapping with an executable statement, the AlamoDE framework cannot report a line number event from a nonexecutable line. A line number event is only reported when a statement in that line number is fetched to be executed. This is reached by checking the assertion source code location before confirming that the assertion is inserted successfully. It checks whether the line number is empty or it is commented out.

Finally, if a referenced variable is an object or a data structure such as a list, this can cause two problems. First, the object is subject to changes under other names because of aliasing. Second, if the object is local, it may get disposed by the garbage collector before the evaluation time. The implementation could be extended to implement trapped variables that would allow us to watch an element of a structure or utilize an aliasing tracing mechanism to retain all changes that may occur under different names. The implementation of temporal assertions presented in this paper does not go after heap variables, which is left for future work.

3.8 Conclusion

DTA assertions bring an extended version of in-code assertion techniques, found in mainstream languages such as C/C++, Java, and C#, into a source-level debugging session. These temporal assertions help users test and validate different relationships across different states of the execution. Furthermore, assertion evaluation actions such as `show` provide the sense of debugging and tracing using print statements from within the source-level debugging session. They give the user a chance to know that the execution has reached that point and the asserted expression evaluated to `true`; it also gives the user the ability to interrupt and stop the execution for more investigation. The ability to log the assertion evaluation result provides the user with the ability to review the evaluation process. A user can check a summary result of what went wrong and what was just fine.

Source-level debuggers provide the ability to conditionally stop the execution through different breakpoints and watchpoints. At each stop, a user will manually investigate the execution by navigating the call stack and variable values. Source-level debuggers require a user to come up with assumptions about the bug and let him/her manually investigate those assumptions through breakpoints, watchpoints, single stepping, and printing. In contrast, DTA assertions require the user to come up with logical expressions that assert execution properties related to a bug's revealed behavior and the debugger will validate these assertions. Asserted expressions can reference execution properties from different execution states, scopes, and over various temporal intervals. Furthermore, unlike conditional breakpoints and watchpoints, which only evaluate the current state, DTA assertions are capable of referencing variables that are not accessible at evaluation time (not active in the current execution state).

DTA assertions do not replace traditional breakpoints or watchpoints, but they offer a technique to reduce their number and improve the overall investigation process. DTA assertions reduce the amount of manual investigation of the execution state such as the number of times a buggy program has to stop for investigation.

Finally, debugging with temporal assertions is not new. In 2002 Jozsef Kovacs et al. has integrated Temporal Assertions into a parallel debugger to debug parallel programs [12]. In 2005 Volker Stolz et al. used LTL over AspectJ pointcuts to validate properties during program execution that are triggered by aspects [11]. In 2008 Cemal Yilmaz et al. presented an automatic fault localization technique using time spectra as abstractions for program execution [13]. However, to the best of our knowledge, we are the first to extend a typical source level debugger's features of conditional breakpoints and watchpoints with commands based on temporal assertion that capture and validate a sequence of execution states (temporal states and temporal intervals). Furthermore, these assertions can reference out-of-scope variables, which may not be live in the execution state at evaluation time.

References

1. Al-Sharif Z, Jeffery C (2009) UDB: an agent-oriented source level debugger. Int J Softw Eng 2(3):113–134
2. Al-Sharif Z, Jeffery C (2009) Language support for event-based debugging. In: Proceedings of the 21st international conference on software engineering and knowledge engineering (SEKE 2009), Boston, July 1–3, 2009, pp 392–399
3. Al-Sharif Z, Jeffery C (2009) A multi-agent debugging extension architecture. In: Proceedings of the 21st international conference on software engineering and knowledge engineering (SEKE 2009), Boston, July 1–3, 2009, pp 194–199
4. Al-Sharif Z, Jeffery C (2009) An agent-oriented source-level debugger on top of a monitoring framework. In: Proceedings of the 2009 sixth international conference on information technology: new generations, vol 00, April 27–29, 2009. ITNG. IEEE Computer Society, pp 241–247. doi:10.1109/ITNG.2009.305

5. Al-Sharif Z, Jeffery C (2009) An extensible source-level debugger. In: Proceedings of the 2009 ACM symposium on applied computing, Honolulu, Hawaii. SAC '09. ACM, New York, NY, pp 543–544. doi:10.1145/1529282.1529397
6. Boothe B (2000) Efficient algorithms for bidirectional debugging. In: Proceedings of the ACM SIGPLAN 2000 conference on programming language design and implementation, Vancouver, British Columbia, Canada, June 18–21, 2000. PLDI '00. ACM, New York, NY, pp 299–310
7. Drusinsky D, Michael B, Shing M (2008) A framework for computer-aided validation. Innov Syst Softw Eng 4(2):161–168
8. Drusinsky D, Shing M (2003) Monitoring temporal logic specifications combined with time series constraints. J Univ Comput Sci 9(11):1261–1276. http://www.jucs.org/jucs_9_11/monitoring_temporal_logic_specification
9. Drusinsky D, Shing M, Demir K (2005) Test-time, run-time, and simulation-time temporal assertions in RSP. In: Proceedings of the 16th IEEE international workshop on rapid system prototyping, June 08–10, 2005. RSP. IEEE Computer Society, Washington, DC, pp 105–110. doi:10.1109/RSP.2005.50
10. Koymans R (1990) Specifying real-time properties with metric temporal logic. Real-Time Systems vol 2, no 4, Oct 1990. Kluwer Academic Publishers, Norwell, MA, USA, pp 255–299. doi:10.1007/BF01995674
11. Volker S, Eric B (2006) Temporal assertions using aspect. Electron Notes Theory Comput Sci 144(4):109–124
12. Jozsef K, Gabor K, Robert L, Wolfgang S (2002) Integrating temporal assertions into a parallel debugger. In: Euro-Par'02, Monien B, Feldmann R (eds) Proceedings of the 8th international euro-par conference on parallel processing. Springer, London, UK, pp 113–120
13. Yilmaz C, Paradkar A, Williams C (2008) Time will tell: fault localization using time spectra. In: Proceedings of the 30th international conference on software engineering (ICSE '08). ACM, New York, NY, USA, pp 81–90. doi:10.1145/1368088.1368100
14. Al-Sharif ZA, Jeffery CL, Said MH (2014) Debugging with dynamic temporal assertions. In: IEEE international symposium on software reliability engineering workshops (ISSREW), 3–6 Nov 2014, pp 257–262. doi:10.1109/ISSREW.2014.60

Chapter 4
Automated Reproduction and Analysis of Bugs in Embedded Software

Hanno Eichelberger, Thomas Kropf, Jürgen Ruf and Wolfgang Rosenstiel

4.1 Introduction

The importance of embedded software increases every year. For example, modern cars currently contain about 100 million lines of source code in embedded software [8]. The highest safety level of ISO26262 standards demands 10^9 h of operation without failure. However, projects in history show that, even with comprehensive testing, bugs remain undetected for years. The control computer of the space shuttle with just 500000 lines of source code was tested overall 8 years with an effort of \$1000 per source code line, i.e., with a total effort of \$500 million [18]. However, it was expected that one bug per 2000 lines of code remained in the last release in 1990. Such bugs may occur in very rare cases and may be only detected while testing the embedded system in a real operation environment.

Static analysis is effectively used during early testing, e.g., unit testing [4]. However, static analysis of bugs is getting close to its limits for complex software. The state space or control flow of big software is difficult to explore completely. Thus, the analysis has drawbacks in performance or in precision. Furthermore, semantic bugs are very application-specific and a wrong behavior cannot be detected even with optimized static analysis tools. Things are getting more severe if a complete and correct specification is not available as golden reference.

System testing describes the process of deploying and testing the software on the target platform. Compared to unit or component testing, which only tests single modules, the software is executed and tested with all integrated software and hardware modules. About 60% of the bugs are not detected before system testing [30].

H. Eichelberger (✉) · T. Kropf · J. Ruf · W. Rosenstiel
University of Tübingen, Tübingen, Germany
e-mail: hanno.eichelberger@uni-tuebingen.de

T. Kropf
e-mail: thomas.kropf@uni-tuebingen.de

J. Ruf
e-mail: juergen.ruf@uni-tuebingen.de

W. Rosenstiel
e-mail: wolfgang.rosenstiel@uni-tuebingen.de

D. Lettnin and M. Winterholer (eds.), *Embedded Software Verification and Debugging*, Embedded Systems, DOI 10.1007/978-1-4614-2266-2_4

However, depending on the software development process model, system testing may be applied only very late in the development process. Real sensors and devices are connected to the embedded system to achieve realistic inputs during system tests. This way, incompatibilities between sensor hardware and the software under test may be detected.

Studies show that the later a bug is detected, the more effort is required to fix it. Some studies present an exponential growth of bug fixing costs during development time [30]. Thus, much more effort is required when bugs are detected during system testing compared to unit testing, caused by close hardware interaction and certification requirements. However, a small percentage of bugs (20%) requires about 60–80% of the fixing effort [11]. When a big portion of complex bugs is detected very late in the development process, project schedules and project deadlines are negatively affected. Thus, the product can often not be placed on the market in time.

One cause for the high effort for bug fixing is the difficulty to reproduce bugs. Users in field or developers during operation tests are often not able to provide enough information to reproduce the bug in the laboratory. Different nondeterministic aspects (e.g., thread scheduling) may make it difficult to reproduce the same execution in the laboratory as the execution observed during operation. About 17% of the bugs of open source desktop and server applications [26] are even not reproducible based on bug reports of the community. The amount may be higher for embedded software with sensor-driven input.

When the bug can be reproduced with a test case, additional effort is required for analyzing the bug. About 8 h are required for running test cases and repairing a bug during system tests [24], if the test case is available. Open source projects like Mozilla receive 300 bug reports a day [5]. With such high bug fixing efforts, it is difficult to handle such high bug rates. Dynamic verification can support the developers in fixing bugs. However, most dynamic analysis tools require the monitoring of the software during runtime. Such monitoring tools are often only applicable on specific platforms.

Bugs are categorized into memory bugs, concurrency bugs, and semantic bugs. Empirical studies show that semantic bugs are the dominant root-cause [37]. The most common semantic bugs are implementations which do not meet the design requirements or which do not behave as expected. Tools to automatically locate root-causes of semantic bugs are required. Memory bugs are not challenging, because many memory profiling tools are available. Concurrency bugs are more problematic, especially their reproduction. More than one out of ten concurrency bugs cannot be reproduced [37].

Our own portable debugger-based approach for bug reproduction and dynamic verification achieves the following contributions to the current state of the art in the area of embedded software development:

- It avoids effortful manual bug reconstruction on any embedded platform by automatically recording and reproducing bugs using debugger tools.
- It improves multi-threaded bug detection by forcing randomized thread switching.

- It decreases the manual debugging effort by applying automated root-cause analysis on reproduced bugs.
- It avoids expensive monitoring hardware by implementing performance optimizations with hierarchical analysis using low-cost debugging tools.
- It reduces porting costs of dynamic verification tools by implementing them in extendable and easily adaptable modules.

Our approach supports developers to detect and reconstruct (mainly semantic and concurrency) bugs faster. It further helps them to fix bugs faster by implementing automated root-cause analyses. Our tool saves debugging costs and hardware investment costs. It is supported by most embedded platforms. It assists developer teams to place their software products earlier on the market.

Section 4.2 gives a brief overview of the normal manual debugging process. Section 4.3 presents methods for automated reproduction of bugs, followed by our own debugger-based approach. Section 4.4 shows how assertion-based verification can be implemented using debugger tools. Section 4.5 presents concepts for analyzing the root-causes of bugs without assertions and concepts for the acceleration of monitoring implementations with cheap debugger interfaces.

4.2 Overview

The workflow in Fig. 4.1 presents the process of manually locating and fixing a bug. It starts by testing the embedded system and the included software during operation in a system testing environment. For example, a navigation software can be executed in a test drive connected to real sensors. During the execution, inputs are being traced and logged into a bug report (A.). This bug report is submitted to a bug report repository. In the laboratory, the developers try to manually reproduce the bug based on the bug report (B.). If the bug can be reproduced, the developers have to manually locate the root-cause of the failure in the source code (C.). After the bug is fixed, the software can be executed with a regression test suite (D.). This way, it is possible to ensure that no new bugs are added during the fixing process.

As presented in Sect. 4.1, all steps of the workflow are usually time-consuming. Our approach presented in the following sections shows that most steps can be supported by automated tools. Automated tools to support bug reproduction (B.) and

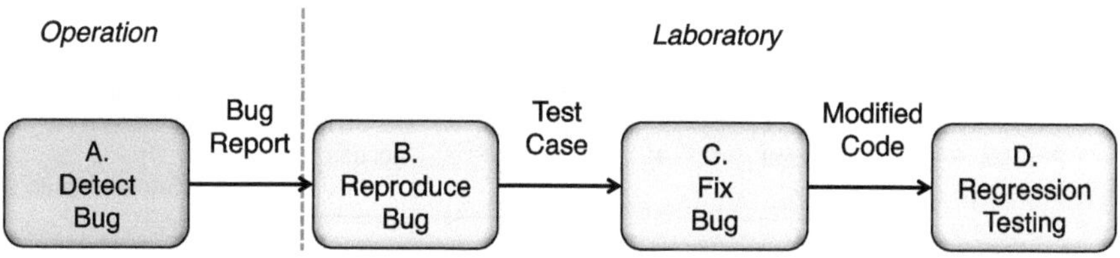

Fig. 4.1 Overview of manual debugging

to generate regression tests (D.) are presented in Sect. 4.3. The bug detection of multi-threaded bugs (A.) is supported by tools presented in Sect. 4.3 as well. Automated bug analyses to support manual debugging (C.) are presented in Sects. 4.4 and 4.5.

4.3 Debugger-Based Bug Reproduction

This section presents approaches of tools for the automated support of bug reproduction. For reproducing bugs, the sensor inputs have to be captured during operation. During replay, the same sensor inputs (e.g., from a GPS or touch screen) have to be triggered or injected to achieve the same execution or instruction sequence. The normal execution of software is deterministic. The same instructions are executed when a software runs with the same inputs. However, these inputs are nondeterministic and are not exactly the same when executing a software twice. For example, it is difficult to achieve the same movements on a touch screen between two test executions. Minimal differences between executions can be determinant whether a failure is triggered. Figure 4.2 shows the different inputs of the software under test (SUT). Zeller [41] lists the nondeterministic inputs of software which are described below. Sensors are the most common input source for embedded software (e.g., from a GPS sensor). User interactions are triggered with human interface devices (e.g., a touch screen). Static data may be stored and read from a disk or flash memory (e.g., an XML configuration file). Access to time or randomness functions in the hardware can change the execution and thus make reproduction difficult. Network interactions may be used for communicating with other devices in the embedded system (e.g., on a CAN bus). The operating environment may be represented by an operating system which controls schedules and memory management. Physical effects may cause hardware changes and they are the most difficult divergence to handle. In the perspective of the SUT, the sources of nondeterminism are categorized into [34]: OS SDK accesses (like system calls), signals or interrupts, specific processor functions, schedules or shared-memory access orderings as well as memory initializations and memory allocations (presented in Fig. 4.2 in the box around the SUT box).

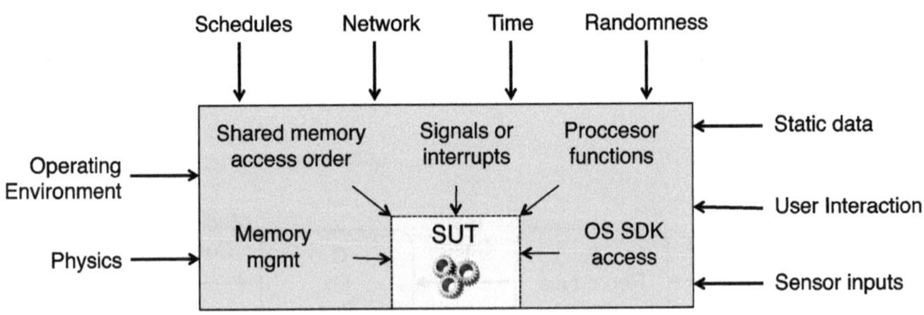

Fig. 4.2 Types of inputs to embedded software

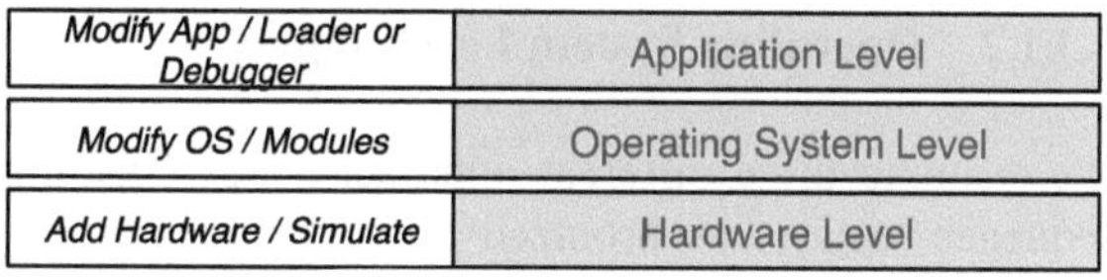

Fig. 4.3 Different levels of replay modules

4.3.1 State of the Art

This section presents current state-of-the-art approaches for the automated reproduction of bugs. The tracing and replaying of the different events caused by the presented sources of nondeterminism can be implemented on different levels (see Fig. 4.3).

It is possible to record/replay the execution of the software on hardware level (Sect. 4.3.1.1). Therefore, additional special hardware is added to support tracing and replay. Another option is to simulate the hardware. In the simulation platform, capturing modules can be integrated. The operating system has control between software and hardware and can record/replay events as well (Sect. 4.3.1.2). It is possible to modify the operating system and to install it on the target platform. Some operating systems provide support to integrate new modules. Furthermore, the software can be modified to record/replay events on application level (Sect. 4.3.1.3). Therefore, the application is modified on source code or binary code level. Debugger tools provide a layer between software and OS and can therefore record/replay events, like presented in Sects. 4.3.2 and 4.3.3. The different approaches are presented in the following sections according to every level.

4.3.1.1 Hardware Level Replay

We examine three types of approaches in the area of hardware level replay: hardware-supported replay, full circuit replay, and virtualization-based replay. Most **hardware-supported replay** approaches consider multiprocessor platforms [21]. To achieve similar executions, the access to shared memory between different cores is recorded and brought into the same sequence during replay. The tracing of shared-memory accesses can be implemented with additional hardware. The advantage of this method compared to software-based approaches is the low overhead achieved by hardware-based functions. **Full circuit replay** approaches add debug instrumentations into the circuits for the FPGA synthesis. This way, it is possible to record and replay the data flow of FPGA circuits in real time [17]. However, only a portion of the program execution can be captured in real time. Other approaches implement **virtualization-based replay** where the hardware is simulated. Moreover, there are approaches for the virtual prototype platform Quick Emulator (QEMU) [10]. However, virtual prototype platforms have a large base overhead which can disturb the interaction with the connected sensor hardware.

4.3.1.2 Operating System Level Replay

We examine three different approaches for operating system level replay: event sequence replay, synchronized event replay, and cycle-accurate event replay. For replaying the same **sequence of events**, some approaches use OS SDK specific commands to trace low-level events and to trigger the same sequence during replay. RERAN [19] uses the *getevent* and an own *sendevent* function of the Android SDK for tracing and replaying on Android mobile platforms. RERAN can record and replay the top 100 Android apps without modifications. Complex gesture inputs on the touch screen can be recorded and replayed with low overhead (about 1%). However, there must be functions available in the OS which support event tracing and sending. Different scheduling or parallel executions on several cores can cause another behavior. Parallel executions require the **synchronization of events** to achieve the same access order to shared resources. SCRIBE [27] is implemented as a Linux kernel module for tracing and injecting. It supports multiprocessor execution and achieves the synchronization of parallel accesses using synchronization points. Using such a synchronization, system calls can be brought into the same order during recording and replaying even when running on multiple cores. However, real-time systems have strict timing requirements and require an **instruction-accurate reproduction of events**, e.g., interrupts. RT-Replayer [33] instruments a real-time operating system kernel to trace interrupts. For replaying, the instructions at memory addresses where interrupts occurred are instrumented with trap instructions (similar to breakpoints of a debugger). Thus, the software stops at traps during replay and the same interrupt functions are triggered.

4.3.1.3 Application Level Replay

For tracing and injecting events on application level, we examine three different approaches for application level replay: source code instrumented replay, binary instrumented replay, and checkpoint-based replay. Using **source code instrumentation**, the source code is modified to trace the control flow and the assignments of variables. Jalangi [35] presents an approach to instrument every assignment of variables and to write the assigned values into a trace file during runtime. During replay, the traced variable values are injected. The approach was presented for JavaScript, but is portable to any other programming language. Drawbacks of Jalangi are: high overhead, big trace files as well as possible side effects of the instrumentation. **Dynamic binary code instrumentation** modifies the source code during execution, e.g., the recording. It dynamically instruments the loaded binary code during runtime, and injects additional tracing code. The instrumented code can be highly optimized and only a low overhead is required for the execution of the code. PinPlay [34] uses dynamic binary code instrumentation with the Pin instrumentation framework. It considers several sources of nondeterminism, as presented in the introduction of this section. However, the Pin framework is only available for specific instruction sets. **Checkpointing approaches** capture the current process state in high frequency (e.g., [40]). Starting at a checkpoint, all nondeterministic events (like system calls) are captured. Checkpoints commonly base on platform-dependent OS SDK operations.

4.3.2 Theory and Algorithms

The previous section presented the way state-of-the-art approaches record and replay bugs on different system levels. This section describes our own approach for tracing and replaying bugs for debugger tools. It considers two sources of nondeterministic inputs: sensor inputs and thread schedules. We do not consider memory violation, because this kind of bug can easily be detected with many available memory profiling tools. We start by presenting the record/replay of sensor accesses followed by the record/replay of thread schedules.

Embedded software often accesses the connected devices triggered by a timer or an interrupt. The device state is requested in a specific frequency. A navigation software may access the GPS sensor with 10 Hz for example. Figure 4.4 shows how a sensor is accessed by a timer function.

The tracing can be implemented by pausing the execution of the software at the location where the access to the device is finished (defined by us as *Receive*). Here, the read data is written to a log file. Figure 4.5 shows this concept.

During the replay the execution is being paused at a location, where the interrupt starts the access to the device (defined by us as *Request*). The access to the sensor is skipped and a jump to the receive location (defined by us as *Receive*) is triggered. At this point, the data is read from the log file. These data are injected into the execution. This way, the sensor data from the recording run is replayed. This concept can be implemented with a debugger tool, as presented in Sect. 4.3.3. The debugger-based replay is illustrated in the sequence diagram in Fig. 4.6.

Other sources of nondeterminism might be thread schedules. The record/replay of thread schedules bases on the reconstruction of sequences of thread events and IO events [28]. This way, the active threads at thread actions, like *sem_wait* and *sem_post*, are monitored. During replay, the same sequences of the invocations of these events are triggered. Listings 1 and 2 show examples of two threads of a

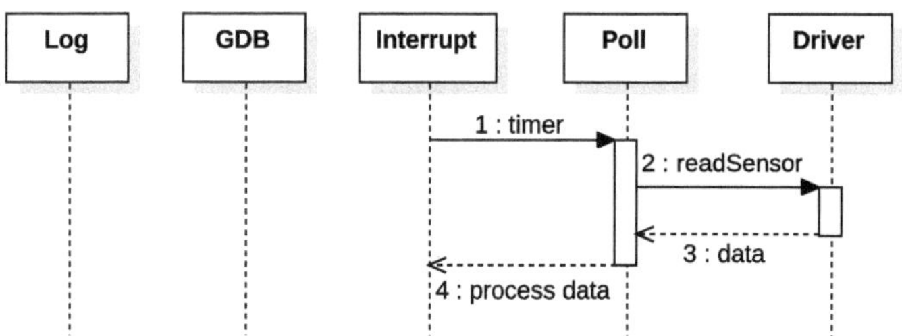

Fig. 4.4 Sequence of sensor accessing

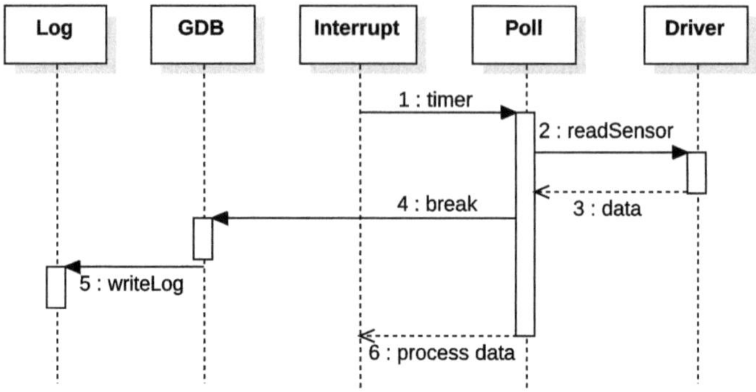

Fig. 4.5 Sequence of recording

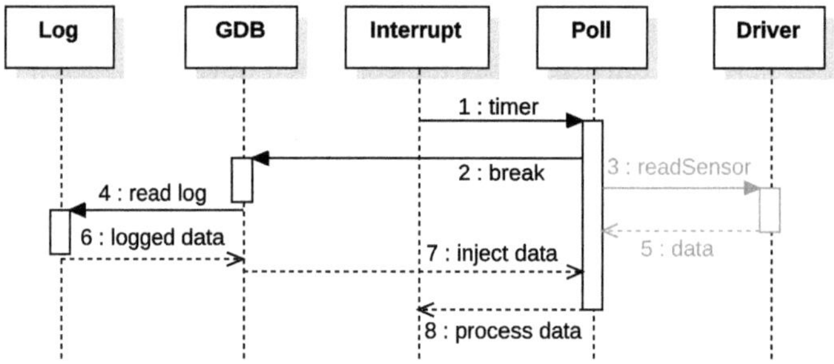

Fig. 4.6 Sequence of replaying

software component for a pedestrian recognition: the thread *Proc* for the recognition of pedestrians in the pictures and the thread *GUI* for drawing rectangles in the pictures where pedestrians were detected. In case the two threads are alternately executed, no failure occurs. However, when thread *Proc* is executed twice, one picture is not drawn. Additionally, when *Proc* is executed twice, the semaphore holds the value two and the *GUI* thread can be executed twice as well. Thus, the current image is released in the thread *GUI* in line 4 and the next call of *drawRec* is triggered with a released image. This case occurs very rarely in normal execution, because, after a long pedestrian recognition in the image in the thread *Proc* (line 2), a thread switch is usually triggered to *GUI*.

Listing 1 Proc

```
1 img=loadImg();
2 recogPed(img);
3 sem_post(sem);
```

Listing 2 GUI

```
1 sem_wait(sem);
2 drawRec(img);
3 showImg(img);
4 releaseImg(img);
```

Our approach implements the serializing of active threads and the randomized switching to threads at thread actions. Therefore, the normal thread scheduler is being locked and only one thread is active at any time. This way, the active thread cannot

Table 4.1 List of tool triggered thread switch actions

Breakpoint	Script actions
sem_init	Set initialization value to the semaphore
sem_wait	If `sem==0` register thread and switch thread, else continue `sem--`
sem_post	Finish post `sem++` and switch to random thread
thread_start	Add available thread into list
thread_end	Delete thread from list
thread_join	Switch to other threads, until joined ones are finished

be preempted by other threads. In our concept, the thread switches are triggered by our tool at thread events (e.g., *sem_wait*, *sem_post*). The scripts monitor the thread events by pausing on the corresponding functions and triggering the thread switches. At each thread event, our tool triggers thread switch actions (see Table 4.1).

Our tool holds a counter for every semaphore *sem*. At every occurrence of *sem_wait*, the algorithm checks whether the semaphore is higher than zero and hence may be passed and the semaphore counter *sem* is decreased. If the semaphore counter is zero, the thread is registered as waiting thread and a switch to another (nonwaiting) thread is triggered. At the occurrence of *sem_post*, the post is finished and a switch to a random thread is triggered. During replay, the same random thread switches are invoked by setting the same seed to the random function like during the tracing. Using this approach, thread switches are always triggered by the scripts. Therefore, the scripts have complete control over thread scheduling.

4.3.3 Implementation

Debugger tools are used to control the execution of the SUT. A popular debugging tool is the GNU Debugger [15], which is available on different embedded platforms. Currently[1], the GDB homepage lists 80 host platforms which are supported by the GDB. Additionally, it is adapted to other platforms by different suppliers. During normal use, the GDB is manually controlled by a developer who types commands into a console terminal. Table 4.2 lists the most frequently used GDB commands.

The GDB is delivered with an external API. Using this API, it is possible to take control over the debugger executions and commands using the Python programming language. Listing 3 shows a simple Python script that can be loaded with the GDB. It starts loading the program to test (line 2) and sets a breakpoint on the method *foo*

[1]Status July 2016

Table 4.2 Thread event actions

Command	Action
file	Loads the program to debug
break	Sets a breakpoint at specific method/line, where execution pauses
run	Starts the execution of the program to debug
continue	Runs the program until next breakpoint
jump	Skips the next lines and jump to a specific location in the code
set	Injects or modifies some variable values
where	Prints current halt point

(line 3). It starts by running the program (line 4). The execution pauses at occurrences of calls of *foo*. The script counts the calls to *foo* (lines 6–8).

Listing 3 Counter GDB Python Script

```
1  import gdb
2  gdb.execute("file␣a.out")
3  gdb.execute("break␣foo")
4  gdb.execute("run")
5  counter=0
6  while True:
7      counter=counter+1
8      gdb.execute("continue")
```

This way, the debugger is controlled with scripted logic. Other debuggers provide different APIs to control the execution of the software under test. Even when the debugger only provides a terminal command interface, these terminal commands can be simulated by scripts and the terminal output can be evaluated as well. Our debugger-based approach records and replays the events using this debugger tool API. Listing 4 shows the way we implemented the replay of GPS sensor data for the Navit navigation software [1].

Listing 4 Record/replay of sensor inputs

```
1  gdb.execute("break␣gpsrequest")
2  gdb.execute("break␣gpsreceive")
3  gdb.execute("run")
4  while running:
5      where=gdb.execute("where",to_string=True)
6      if record and "gpsreceive" in where:
7        lat=gdb.execute("print␣lat",to_string=True)
8        ... # Access lng and angle
9        trace.write(lat,lng,angle)
10     elif not record and "gpsrequest" in where:
11       data=trace.read()
12       gdb.execute("jump␣gpsreceive")
13       gdb.execute("set␣%s"%(data.lat))
14       ...# Set lng and angle
15     gdb.execute("cont",to_string=True)
```

A breakpoint is set in the source code lines where the request to a device is started and where the access to the device is finished (lines 1–2). For the recording, the execution pauses at the location where the GPS data was received. In this case, current GPS values are printed (lines 7–8) and written to a log file (line 9). For the replaying, the execution pauses at the location where the access to the GPS is started. The access is skipped with the *jump* command (line 12) and the data from the log is injected (lines 13–14). If the debugger-based recording is too slow using breakpoints, it can be implemented with *printf* statements or a trace buffer implementation.

The following paragraph presents how the program under test can be controlled to achieve deterministic thread schedules. The GDB provides the commands listed in Table 4.3 for debugging multi-threaded programs. It includes the three commands we used for our implementation. At every breakpoint pause, the developer can manually examine the current thread or can switch to other threads that are contained in the thread list.

Listing 5 presents an implementation for the replay of the pedestrian recognition (and is similar for the game case study presented in Sect. 4.3.4).

Table 4.3 List of GDB commands for handling multiple threads

Command	Action
info threads	Shows current loaded threads
thread	Switch to another thread
set scheduler-locking on	This commands disables the normal thread scheduler

Listing 5 Replay for two-threaded pedestrian recognition

```
1  gdb.execute("break␣main")
2  gdb.execute("run")
3  gdb.execute("set␣scheduler-locking␣on")
4  gdb.execute("break␣sem_wait")
5  gdb.execute("break␣sem_post")
6  random.seed(trace.readSeed())
7  while running:
8      where=gdb.execute("where",to_string=True)
9      if "sem_wait" in where:
10             if sem==0:
11                     gdb.execute("thread␣%s"%Proc)
12             else:
13                     sem=sem-1
14      elif "sem_post" in where:
15             sem=sem+1
16             t=random(Proc,Gui)
17             gdb.execute("thread␣%s"%t)
18      # Sensor data replay code
19     ...
```

The command `"set scheduler-locking on"` disables the current thread scheduler (line 5). When this option is activated, only one thread can be executed at any time. Similar effects can be achieved with portable non-preemptive thread libraries [16]. This way, the parallel execution of threads is serialized. For implementing the required thread switches, our tool sets breakpoints at the used thread actions (see lines 4–5). At every passing of a thread action, our tool triggers the thread switches (Listing 5, line 14–17) with the *thread* command of the GDB based on actions presented in Table 4.1. The control of thread schedules is integrated with the sensor replay (Listing 5, after line 18).

Using randomized switching at thread actions achieves a better thread interleaving coverage than the normal thread scheduler. This way, concurrency bugs can be manifested faster.

4.3.4 Experiments

Figure 4.7 shows our measurements [14] for tracing or recording the sensor input data of the single-threaded software Navit executed on Ubuntu Linux on an X86 Intel platform. The measurements consider a route with 1200 GPS coordinates. These coordinates are read from a file by a Mockup GPS server. The GPS frequency was tested at 50, 33, 20, and 10 Hz, and in parallel, the user cursor inputs (e.g., from touch screen) was captured at 100 Hz. The overhead between the normal execution (labeled as *Normal*) and the recorded execution (labeled as *Rec*) remains nearly

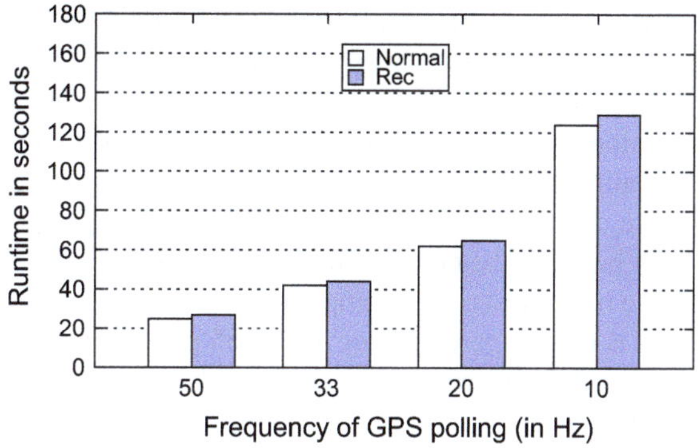

Fig. 4.7 Performance measurements for sensor input recording for single-threaded Navit [14]

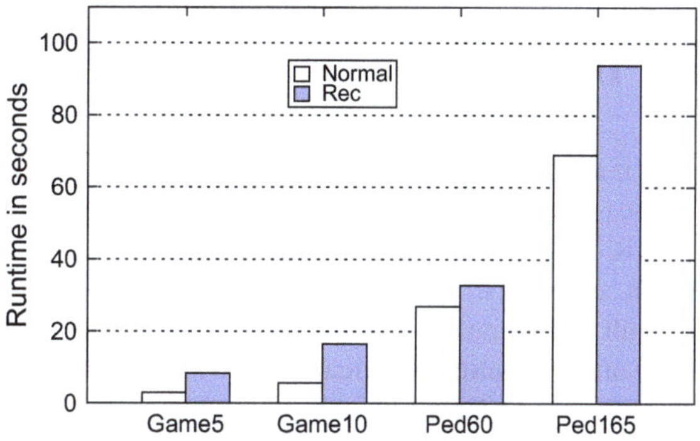

Fig. 4.8 Performance measurements for deterministic scheduling and recording

constant, since the time used to pause the execution at breakpoints is caught up by the invocations of the polling timers. This way, our approach for timer-based software achieves minimal overhead. Tracing optimizations are required for other types of software, e.g., by grouping several inputs and only tracing a set of inputs at once [13].

We tested our approach for deterministic scheduling and recording with two embedded software examples implemented with POSIX threads. The performance measurements for these examples are presented in Fig. 4.8. The first is an ASCII-based fly and shoot game. The game uses two threads, one for drawing the scene and one for reading from the keyboard. The second example is a pedestrian recognition [38] in video data of a vehicle camera (see Sect. 4.3.2). For every example, we used two scenarios for each measurement, a short and a long one. We measured the game until 5 or 10 lives were lost without interaction of the user. We measured the pedestrian recognition with a set of 60 or 165 pictures as inputs. Our experiments were executed five times on an NVIDIA Tegra K1 with ARM CPU and Linux OS.

During recording the scenario of the game, in average 377 thread switches are scheduled for the short scenario and 642 thread switches are scheduled for the longer scenario. For the pedestrian recognition example, in average 186 (short) and 475 (long) thread switches are scheduled. The recording of the pedestrian software requires, in average, 1.22X and 1.36X overhead. The overhead for recording the game is higher with, in average, 2.86X and 2.98X, because more thread switches have to be triggered in a shorter time.

We observed that the overhead keeps similar for short and long scenarios in both case studies. The measurements show that performant recording can be implemented with minimal effort. Every recording and replay script contains less that 50 lines of source code. Additionally, the pedestrian recognition example shows that concurrency bugs can be detected faster when forcing randomized switches with our tool. Hence in our tests, the example bug is detected in a few seconds with our approach, but it is not triggered during 10×165 picture inputs with the normal thread scheduler. For the performance measurements, we triggered the alternate invocation of the two threads for not triggering the bug.

4.4 Dynamic Verification During Replay

Even if a bug can be reproduced, it is difficult to locate the root-cause of the bug during the replay. Manual debugging of a replay (e.g., with GDB) is effortful. The source code is often implemented by other developers. Thus, it is difficult to comprehend which sequence of actions (e.g., method calls) leads to the failure. Additionally, it is difficult to understand how the faulty sequence of actions is caused. The approaches in the area of runtime verification provide concepts for automatically analyzing executions during runtime by comparing them against a formal specification. Runtime verification tests whether a set of specific properties are held during the execution. The components which observe the execution are called monitors.

4.4.1 State of the Art

The *online monitors* approach runs a monitor in parallel to the execution during operation. Online monitors have to be very efficient, because the normal execution should not be disturbed. However, online monitors may react to observed anomalies during the execution [29]. At the occurrence of anomalies, fail-safe or recovery modes may be activated during operation. *Log monitors* examine log files, captured during a long-term execution of a software. The trace files are efficiently generated during operation. The tracing should be lean or implemented with fast additional hardware for not disturbing the normal execution. Offline, the trace file is analyzed in detail [6]. Wrong event sequences in the trace can be detected, pointing to the failure or even to the root-cause of the failure. Some approaches combine record/replay and dynamic analysis of software [32, 40, 41]. However, they do not use a framework for the implementation of complex assertions and were not tested on embedded platforms.

Table 4.4 presents the advantages and disadvantages of each mode. The two approaches do not provide support to check whether the failure still occurs or not (*Control Test*). Moreover, they cannot be applied fine-grained (*Fine-grained*),

Table 4.4 Comparison of different characteristics for the monitoring types

Type	Online	Log	Replay
Control test	✗	✗	✓
Fine-grained	✗	✗	✓
Recovery	✓	✗	✗
Long term	✓	✓	✗

because they would disturb the normal interaction with the user or with other systems. Our replay approach fulfills the first two categories (*Control Test* and *Fine-grained*). However, the activation of *recoveries* is only possible with the online mode as summarized in Table 4.4. *Long-term* tracing and monitoring is, in our opinion, best applicable with tracing logs.

4.4.2 *Theory and Workflow*

The concept of applying dynamic verification during replay [12] is based on the concept of tracing only the relevant inputs to the software and replaying them offline. During replay, fine-granular tracing can be executed. Monitors or analysis can check these traces for anomalies. During replay, the requirements for efficient tracing and monitoring are less compared to normal operation. Additionally, the generated replay can be used later as a control replay, after the bug has been fixed. The replay concept is, in our opinion, the best option for system testing, because control replays which can be used as regression test cases later are generated. Figure 4.9 shows how the different manual steps are supported or replaced by automated tools. The detection of multi-threaded bugs (A.) is optimized by the randomized scheduling concepts presented in Sect. 4.3.3. The replay of bugs (B.) is automized (see Sect. 4.3) and can be used as regression test (D.). Automated analyses during replay (B.) support the manual bug fixing (C.). These analysis tools are presented in the following sections.

A. Systematic Bug Detection: The software is tested in real-world operation. The incoming events to the software are captured during these tests. Recording mecha-

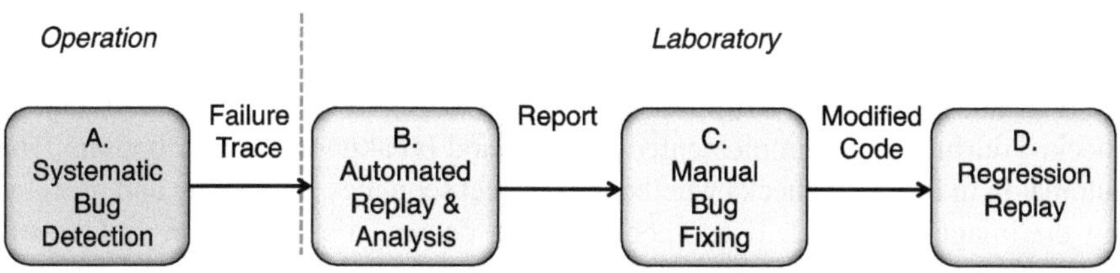

Fig. 4.9 Workflow for debugger-based dynamic verification during replay

nisms are implemented with a symbolic debugger in order to avoid instrumentation and to achieve platform compatibility. Therefore, the debugger is controlled by a script. The developer decides which events are relevant and have to be captured. Thus, the recording can be kept lean. To efficiently detect multi-threaded bugs, the thread scheduler is controlled by our scripts triggering randomized switches at any thread event.

B. Automated Replay and Analysis: The failure sequence can be loaded and deterministically replayed in the laboratory. The replay mechanisms are implemented with portable debugger tools. The software is executed with the debugging interface on the same hardware as during operation for arranging the same system behavior as during the original run. During replay, the failure occurs again based on the deterministic replay. Manually debugging the complete execution sequence or even several processing paths of events is very time-consuming. Therefore, we apply dynamic verification during replay to automatically detect potential anomalies. These information can give a hint to the fault. Analyses performed online during operation disturb the normal execution, but do not cause drawbacks during replay, because no interactions with the user or with external components are required for the execution of the replay.

C. Manual Bug Fixing: Based on the report of the dynamic verification, the developer can manually fix the fault. The step results in a patched program.

D. Regression Replay: The modified program can be tested with a control replay. It is executed with the recorded sequence of failure inputs to observe whether the faulty behavior occurs again. Finally, the bug replay can be archived as a test case for a regression test suite.

4.4.3 Implementation of Assertions During Replay

In our workflow presented in the previous section, dynamic verification is applied during replay for detecting the cause of the bug. This section shows how this cause of a bug can be detected using assertions. Such assertions can be easily implemented with a debugger. Therefore, during debugger-based replay, the execution can be monitored with the debugger as well. The following paragraph considers the multi-threaded replay of a pedestrian recognition software (as presented in Sect. 4.3). During the replay of the software, the event sequence can be analyzed using assertion-based verification. The sequence of method calls is monitored by setting breakpoints on the corresponding methods. In the pedestrian recognition example, three events are relevant: **recogPed()**, **drawRec()**, and **showImg()**. Temporal conditions can be checked during replay, implemented with method breakpoints or watchpoints. The automaton in Fig. 4.10 checks whether the correct sequence for loading and processing the images is called. If another transition occurs, a specification violation is detected.

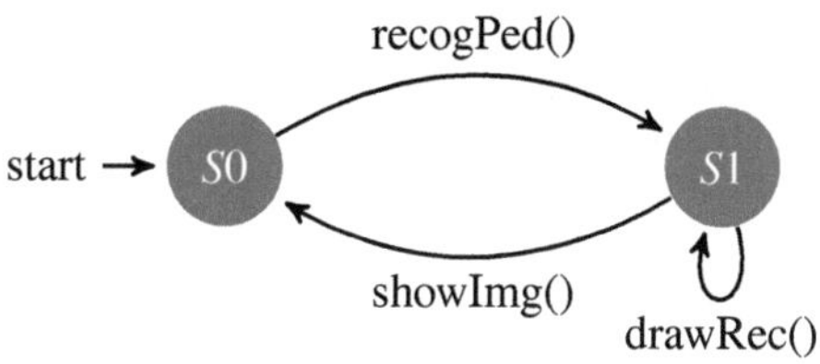

Fig. 4.10 Automaton for the action sequence of the pedestrian recognition

This kind of monitor can easily be implemented in a GDB Python script (see Listing 6). Lines 1–3 set the breakpoints and watchpoints on the presented conditions. Lines 6–14 check the reached point and the current state.

Listing 6 Runtime Verification GDB Python script

```
1  gdb.execute("break␣recogPed")
2  gdb.execute("break␣drawRec")
3  gdb.execute("break␣showImg")
4  state=0
5  while running:
6            where=gdb.execute("where",to_string=True)
7            if "drawRec" in where and state==1:
8                    state=1
9            elif "recogPed" in where and state==0:
10                   state=1
11           elif "showImg" in where and state==1:
12                   state=0
13           else:
14                   print "Spec␣violation!"
...          # Sensor and Thread Replay Code
```

4.4.4 Experiments

Figure 4.11 shows our performance measurements for different monitoring scenarios compared to the multi-threaded recording overhead.

We measured the runtime when 2, 5, and 10% of all methods of the corresponding software were monitored using the GDB (this represents the *online monitor* or *log monitor* mode). The monitored methods can be compared to parallel running state machines as presented in the previous section. We randomly selected a specific percentage (2, 5, or 10%) of methods and set breakpoints on them. We measured the runtime for executing our multi-threaded recording approach (labeled as *Record*) as well. We used the same four scenarios as in Sect. 4.3: the fly and shoot game with 5 or 10 lives and the pedestrian recognition with 60 or 165 video pictures. In all scenarios, the monitoring of 2% of the methods is faster than the recording.

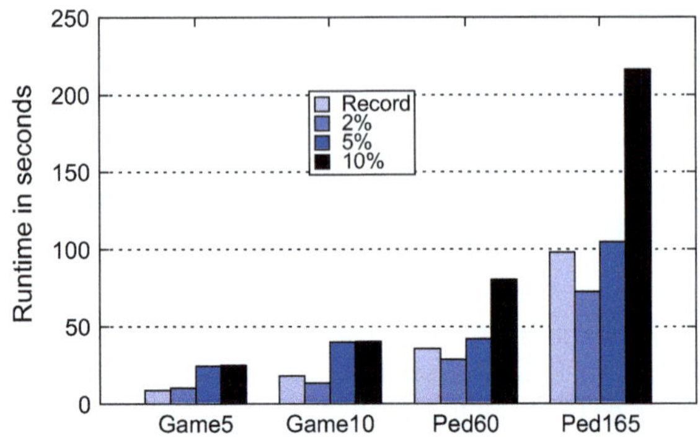

Fig. 4.11 Comparing the overhead for our recording with methods monitoring overhead

Table 4.5 Number of GDB actions for recording and methods monitoring

Scenario	Record	2%	5%	10%
Game5	383	7069	13050	13333
Game10	810	7069	18252	18840
Ped60	181	910	7099	17143
Ped165	496	2117	17546	47593

The monitoring of 5% of the methods is slower than the recording, especially when considering the game scenarios. The recording of the execution of the scenarios is in average 2.45 times faster than the runtime for monitoring of 10% methods. However, the performance of method monitoring mainly depends on how often every method occurs in the execution. Table 4.5 shows the amount of GDB pauses for the recording scenarios and every method monitoring scenario (2, 5 and 10%).

The higher the number of pauses, the higher the recording or monitoring time. However, the recording had a smaller number of pauses compared to the monitoring scenarios, the proportional runtime overhead was higher. The reason for this finding is that more GDB commands are required at every breakpoint pause for implementing the recording. Additionally, we tested the monitoring of 15% of methods of the pedestrian recognition, but the monitoring of the first picture was not finished after 10 min.

4.5 Root-Cause Analyses

In the previous section, we showed how wrong action sequences can be detected by comparing the executed actions with a specification property. The bug was triggered by the wrong sequence causing the program to crash. However, in many cases, a correct or complete specification of actions is not available. Usually, this is the case, as a specification property often already points to a potential failure, which can be fixed manually. In this section, we consider semantic bugs (in software without a specification), i.e., the root-cause of the bug is found in the value processing or program logic. We present concepts to detect wrong or missing method calls in processing and to identify the root-cause of the corresponding error logic.

4.5.1 State of the Art

This section presents the state of the art in the area of fault localization of noncrashing bugs and embedded software monitoring.

4.5.1.1 Delta Debugging

The book 'Why Programs Fail' [41] presents different concepts for fault localization in software. It describes several concepts for dynamic analyses, including **delta debugging** and **anomaly detection**. Some of these concepts are similarly considered in our work, e.g., for our delta computation approach. Later work of Burger and Zeller [7] developed **dynamic slicing** for the localization of noncrashing bugs. They apply several steps to isolate the failure location by following back the bug in the execution. However, delta debugging bases on experiments with the program to automatically generate passing runs and failing runs, which is difficult and runtime-consuming in embedded contexts (like mentioned by [3]).

4.5.1.2 Dynamic Verification for Noncrashing Bugs

Zhang et al. [42] implement an approach to detect noncrashing bugs caused by wrong configurations. It profiles the execution of failing and not failing configurations. Many embedded softwares do not even require a configuration and the bugs can be located in the source code. Liu et al. [31] apply **support vector machines** to categorize passing runs and failing runs of noncrashing bugs. Their approach generates behavior graphs on method level to compare different runs. For classification, a lot of input runs are required and only suspect methods can be detected (not the relevant source code lines). Abreu [2] implement fault localization for embedded software using **spectrum-based coverage analyses**. They apply model-based diagnosis to improve the results of the analyses. Similar to Tarantula [25], the coverage of executed statements of each failing run is compared to the passing runs. They assume that a big set of test cases of failing and passing runs are available. However, all approaches require a set of failing runs, which can be used for classification. In our use case for system testing, a big set of nonfailure runs and failure runs is not available for classification.

4.5.1.3 Monitoring of Embedded Software

For the implementation of fault localization, the software has to be monitored. Amiar et al. [3] use special **tracing hardware** to monitor embedded software. They apply spectrum-based coverage analyses on a single trace. When detecting a failing cycle, it is compared to previous similar ones to detect spectrum-based coverage deltas. However, they assume a tracing hardware for the specific embedded platform is available. Such hardware is usually expensive. Several runtime verification approaches [20, 36] use **cheap debugger interfaces** with the GNU Debugger (GDB) [15] to achieve platform compatibility, but they do not present a concept to detect bugs without a specification. FLOMA [23] observes the software fine-grained using **probabilistic sampling**, but it does not monitor every source code line. It randomly decides if a specific execution step is monitored. However, probabilistic sampling can miss important steps and FLOMA requires the instrumentation of the source code. Zuo et al. [43] present a **hierarchical instrumentation** approach to accelerate monitoring. Their approach instruments the software to monitor and analyze the method call sequences. Afterwards, only the suspect parts are monitored on source code line level. This way, the monitoring can be accelerated. Our approach extends this approach and applies it to debugger tools.

4.5.2 Theory and Concepts

We apply root-cause analysis on a failure replay to automatically detect suspect source code lines which are potential root-causes of the failure (based on [3, 13, 14]). This analysis results in a report, which can give the developer a hint where the bug is located in the source code. In the following, we present a workflow which bases on a failure replay and a nonfailure replay. We split the execution of the software into parts (see Sect. 4.5.2.1). Several executions of a partition have overlaps and can be compared. Every execution of one partition is called a run. Afterwards, this run in the replay which executes the failure has to be detected (see Sect. 4.5.2.2). The failure run in the replay is compared to several runs in the replay which are similar and which are not categorized as failing runs (see Sect. 4.5.2.3). For the comparison of the failure run to the similar runs, we apply fine-granular analyses on source line level, aiming at detecting the buggy source code line. We show metrics for coverage analysis as well as invariant generation analysis (see Sect. 4.5.2.4). However, fine-grained monitoring can be very slow using cheap debugger interfaces. Therefore, we present an acceleration approach in Sect. 4.5.2.5.

We exemplify our own approach with a noncrashing bug in the open source navigation software Navit [1]. If Navit receives GPS sensor data with an angle smaller than −360, the vehicle pointer is not drawn for a short amount of time. This bug might disturb the driver, e.g., when looking for the correct crossing on a busy street. This bug does not throw an exception. It can just briefly be observed in the GUI.

It is caused by the processing of wrong sensor data (the sensor sends angle values <-360). Such a case occurs when the software is not compatible with the sensor hardware outputs. The bug is caused by a wrong calculation in the Navit software (presented in the following).

4.5.2.1 Partition Replay into Runs

The approaches of the state of the art in the area of anomaly detection provide concepts for detecting root-causes by comparing nonfailing with failing execution runs of the software under test [41]. However, the execution of complex embedded software may contain parts which execute different functionalites. The comparison of different functionalities can cause many false positives, especially when only a small set of reference runs is available. In our approach, we split the execution of the embedded software in several comparable parts (executing similar functionalities). Embedded software usually processes sensor data to update the program states. This processing is usually very similar every time it is executed. Figure 4.12 exemplarily presents the concept for the replay of Navit. A replay of Navit contains different types of processing, e.g., GPS processing, touch screen input processing or traffic data processing.

The processing of sensor data starts after the data has been read from the sensor hardware (in Sect. 4.3 defined as the point *receive*). A processing is represented by the following tuple:

$$Processing = (Start, Run, End) \tag{4.1}$$

Start and *End* represent source code lines passed in the execution. *Start* is the position in the execution where the system starts to process this sensor data. *End* is the position in the execution where the processing of the sensor data is finished. The processing *Run* includes a list of all operations *Ops* and method calls *M* in the processing of the sensor data:

$$Run = (M, Ops) \tag{4.2}$$

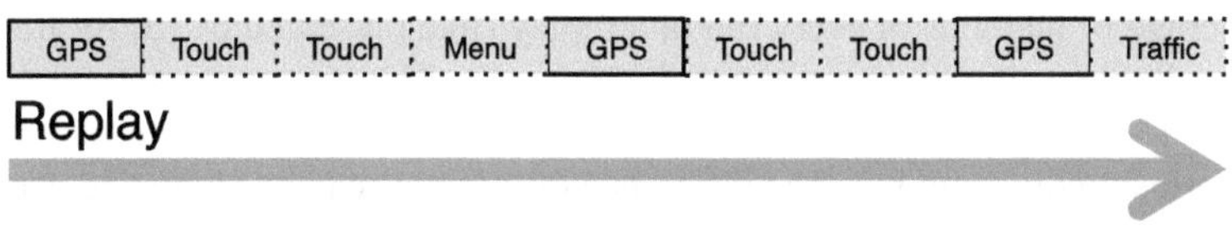

Fig. 4.12 Partitioning of a replay into runs

and

$$M = \{m_1, m_2, \ldots, m_{N1}\} \quad (4.3)$$
$$Ops = \{op_1, op_2, \ldots, op_{N2}\} \quad (4.4)$$
$$op_x = \{v_1, v_2, \ldots, v_{N3}\} \quad (4.5)$$

In our approach, the execution breaks at *Start*. Beginning at this breakpoint, the processing is observed either on method or on source code line level. On method level, every method $m_i \in M$ is monitored (4.3). On source code line level, every operation $op_i \in Op$ is monitored (4.4). The monitoring stops with the execution of *End*. We consider one statement or source code line as operation op_i. Each operation op_x holds several global, local, and argument variables $v_1, v_2, \ldots, v_{N3}$ (4.5). The runs of the same type can be compared by detecting differences considering the executed methods, the executed operations or the observed variable values. In our current approach, *Start* and *End* have to be specified with debugger scripts. Other approaches implement automated cycle detection [3], which can be similarly applied to our approach.

4.5.2.2 Detect the Failure Run

Every execution of the partition of the software is considered a run of a replay. When running the failure replay, one or more runs of a specific sensor processing cause the observed failure. We present a lightweight concept for the detection of the failure run in the replay. It classifies those runs as failing which are most different to the runs in a nonfailure replay (as described in the following). Every run of a replay can be compared to other runs, because similar operations and methods are executed. Differences in the run may point to the failure. To detect the failure run, our approach expects two replays. One replay which causes the failure and a second replay which does not cause the failure. The runs of a replay with a failure can be compared to the runs of a replay not causing an observed failure. Two runs can be compared by checking which source code lines or methods are covered by a run. As being presented in Sect. 4.5.2.5, it is more efficient to consider the coverage of methods in this stage.

Table 4.6 shows an example matrix (representing the Navit bug), which contains the coverage of methods of every run of a replay (like presented by [3] for traces).

The callable methods are represented in the rows. The runs are represented in the columns. The value 1 in a cell means that the method in this row is executed by the run in the corresponding column. The difference of two runs can be compared using the hamming distance, i.e., by counting the differences in rows between the two columns of runs. In our example: $distance(Run1, Run7) = 2$, $distance(Run2, Run7) = 3$ and $distance(Run3, Run7) = 3$. For a straightforward presentation, we consider some pseudo-methods $m1 - m4$ and the *draw* method.

Table 4.6 Coverage matrix for monitored runs of a replay

Methd/run	Run_1	Run_2	Run_3	...	Run_7
m_1	1	1	1	...	1
m_2	1	0	1	...	1
m_3	1	1	0	...	1
m_4	1	1	1	...	0
draw	1	1	1	...	0
...	...	...	...	...	...

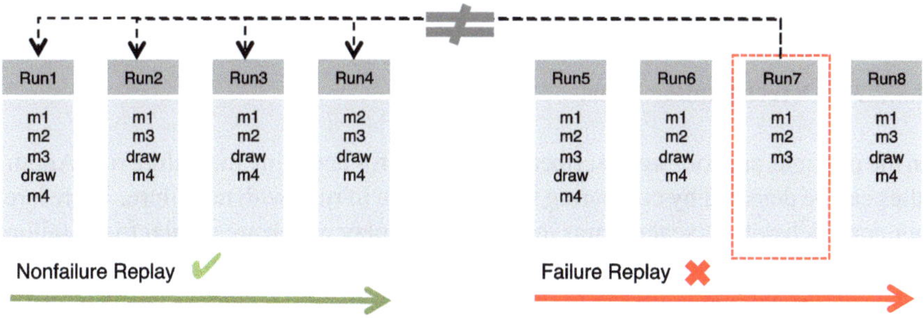

Fig. 4.13 Detect the failure run

The failure is detected by comparing every run in the failure replay with every run in the nonfailure replay using the matrix presented above and the hamming distance. The run in the failure replay which is not similar (or most different) to all runs in nonfailure replay is considered a failure run. Figure 4.13 shows how *Run*7 in the failure replay is compared to every run in the nonfailure replay. *Run*7 is most different to the nonfailure runs, because the call of the *draw* method of the vehicle pointer is missing. The number of occurrences of every method can be combined with the hamming distance as well. Differences between two cells higher than a threshold *threshold* can be counted as 1 and else as 0. This way, one method of two runs can be categorized as different if one run executes a method many more times than the other run. It is possible to consider the call sequence as well. However, the method coverage can be monitored faster than source code line coverage (see Sect. 4.5.4). Additionally, it is possible to categorize several runs as failing runs. For example, it is possible to categorize those ones as failure runs which show 20% differences when compared to the runs of a nonfailure replay. However, the following explanations base on one failure run. Note: If several runs are ranked with same distance, the latest is chosen (because the bug is expected to be near the end of the replay).

Here, *Run*7 in the failure replay is most different to all other runs in the nonfailure replay. That means: *Run*5, *Run*6, and *Run*8 have a corresponding run in the nonfailure replay with a hamming distance, which is smaller than the hamming distances of *Run*7 to each run of the nonfailure replay.

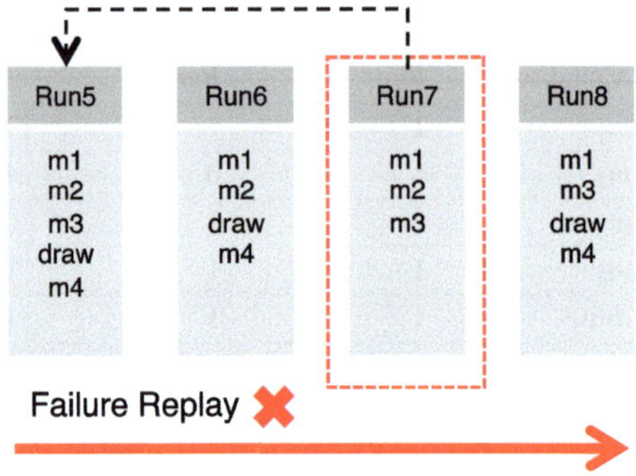

Fig. 4.14 Detect the similar runs

4.5.2.3 Detect Similar Runs

In the previous section, we described the concept for detecting the failure run. Anomalies can be detected by comparing this failure run to runs with no failure. Therefore, our approach detects several runs in the failure replay which are similar to the failure run, but which are not categorized as failing runs, i.e., it detects those runs which have similar method coverage to the failing run using the hamming distance (bases on [3]). Figure 4.14 shows the failure replay. In this example, the most similar run to the failure run *Run*7 is *Run*5. This way, the failure run is compared in detail to the runs which are most similar to it. This concept bases on the nearest neighbor model, which was similarly applied to log files of a tracing hardware [3]. Our approach detects similar runs in the same replay where the failure run occurs at, these similar runs being executed under the same context as the failure run (e.g., considering the configuration context).

In our tests, we detected three runs which are similar to the failure run. These runs and the failing run are compared in detail using delta analysis.

4.5.2.4 Delta Computation

The failing run and the similar runs are compared in detail to detect deltas which can point to the failure root-cause. When comparing the similar runs to the failing run in the failure replay, different metrics can be applied to identify suspect source code lines. We present the concepts and metrics for the delta analyses based on the Navit bug example.

The Navit bug is based on a wrong calculation in the GPS processing (pseudocode presented in Listing 7). If the angle is smaller than 0, the value 360 is added to the angle in line 2. However, in the case the angle is smaller than -360, the angle keeps a negative value after line 2 and lines $5 + 6$ are skipped and the vehicle pointer is not drawn. In a correct implementation, a mathematical modulo operation should be applied to the angle computation to generate a positive value. Line 5 and the parameter variable *lazy* are explained in the following section.

In our example, the following GPS input sequence to the *vehicle_update* method triggers the bug $(\ldots\{42, 9, 40, 0\}, \{42, 9, 55, 0\}, \{42, 9, \boxed{-370}, 0\}, \{42, 9, 70, 0\})$. Here, the third input sends wrong angle data, e.g., from a sensor device.

Listing 7 Navit bug example

```
0 void vehicle_update(latitude,longitude,angle,lazy)
1         if (angle<0)
2                 angle+=360;
3         ...
4         if (angle in range){
5                 setDrawingMode(lazy);
6                 draw();
7         }
8         return
9 }
```

We apply fault localization metrics to detect the root-cause of such bugs in the failure replay (in the example, the wrong calculation in line 2). We define these metrics with three coefficients, which are generated for every executable source code line. These coefficients count the number of runs which fulfill a specific characteristic for a specific source code line *op*. These characteristics define the amount of runs causing a failure or not failure. And they define whether they cover the specific source code line *op* or not.

- e_p: #runs with **nonfailure**, which *execute the line.*
- e_f: #runs with **failure**, which *execute the line.*
- n_p: #runs with **nonfailure**, which do *not execute the line.*
- n_f: #runs with **failure**, which do *not execute the line.*

Popular metrics for fault localization were given by Tarantula, Jaccard, and Occhiai. The metrics are defined as follows [39]:

$$Tarantula : d_T = \frac{\frac{e_f}{e_f+n_f}}{\frac{e_f}{e_f+n_f} + \frac{e_p}{e_p+n_p}} \tag{4.6}$$

$$Jaccard : d_J = \frac{e_f}{e_f + e_p + n_f} \tag{4.7}$$

$$Occhiai : d_O = \frac{e_f}{\sqrt{(e_f + e_p)(e_f + n_f)}} \tag{4.8}$$

Tarantula measures which lines are primarily executed by failing runs. These lines are considered as more likely to be the root-cause of the failure. The suspicious ranking is decreased if many nonfailing runs execute the line as well. The *Jaccard* coefficient is based on e_f as well, but results in a less suspicious ranking when many nonfailing runs execute the line or many failing runs do not execute the line. *Occhiai* additionally weights the difference between e_p and n_f. Thus, the ranking is lower, when many nonfailing runs execute the line and many failing runs do not execute the line at the same time. The previously presented metrics mainly consider, which source code lines are often represented in the failing runs. However, in case of noncrashing bugs of embedded software, the error of the bug may be the missing call to an OS SDK library. Additionally, missed source code lines in the failing run may point to the wrong evaluation of conditional logic. A comparison of missing code in the failing run compared to the nonfailing run can point to the failure. Thus, in our opinion, it is required to rank missing features in the failing runs as well. The AMPLE metrics (4.9) consider missing features in the failing run as well [9].

$$AMPLE : d_A = \left\| \frac{e_f}{e_f + n_f} - \frac{e_p}{e_p + n_p} \right\| \tag{4.9}$$

Table 4.7 shows the different results of the metrics for the presented Navit bug example (see Listing 7) for three similar runs (Run_{Sim}) and one failing run (Run_{Fail}). The higher the resulting factor, the higher the suspiciousness of the corresponding source code line. We observed that every metric ranks the operation with index 2 as suspect. However, operations 5 and 6, which additionally point to the wrong conditional case, are not ranked as suspect, despite by AMPLE.

In most fault localization approaches, the metrics result in a list of source code lines with their according suspicious rankings. However, it is difficult to evaluate the source code manually based on this list. In our use case, our tool sets breakpoints on the most suspect source code lines during replay. The developer can step through the suspect lines during replay and check which lines are executed in the failing and the nonfailing runs. Our tool only sets breakpoints on source code lines with AMPLE suspicious ranking 1. Additionally, it notes at every suspect source code

Table 4.7 Calculation of metrics by example

Methd/op	op_1	op_2	op_3	op_4	op_5	op_6
Run_{Sim}	1	0	1	1	1	1
Run_{Sim}	1	0	1	1	1	1
Run_{Sim}	1	0	1	1	1	1
Run_{Fail}	1	1	1	1	0	0
e_p	3	0	3	3	3	3
e_f	1	1	1	1	0	0
n_p	0	3	0	0	0	0
n_f	0	0	0	0	1	1
Tarantula	0.5	1	0.5	0.5	0	0
Jaccard	0.25	1	0.25	0.25	0	0
Occhiai	0.5	1	0.5	0.5	0	0
Ample	0	1	0	0	1	1

line, whether it is executed in the failure run (but not in any similar run) or whether it is executed in every similar run (but not in the failure run). Other approaches propose the combination of suspicious metrics [39], e.g., combining AMPLE and Occhiai. This way, machine learning algorithms generate weighted combinations of different metrics. Studies show that better metric results are achieved with combined metrics. However, a learning phase which is not possible in our use case is required.

Previously, we showed how delta computation is able to show coverage deltas between the failure run in the failure replay and some similar runs in the replay. However, root-causes which start to propagate at a wrong variable assignment can often only be detected by monitoring all variable values in every source code line.

In our example, another sequence of GPS inputs may trigger the bug: ({42, 9, −340, 0}, {42, 9, −355, 0}, {42, 9, $\boxed{-370}$, 0}, {42, 9, −355, 0}). This sequence can happen, when the sensor sends data smaller than 0 and incrementally switches to an angle smaller than −360. Applying coverage-based analyses during the replay of this sequence, the wrong source code line (line 2 in Listing 7) cannot be detected, because line 2 is executed in every run. However, the root-cause can be detected when monitoring all variable values in every source code line.

Anomalies in variable values can be detected using invariants. Invariants are characteristics of variable/value pairs which are stored for each run. The invariants being held by the nonfailing run can be compared with invariants stored from the failing run. Therefore, we automatically generate invariants for the nonfailing and failing run. Range invariants check, which ranges of variable values are observed during a run. In our Navit example, the invariant in Eq. (4.10) is stored by every nonfailure run. This invariant can be checked in the failure run, where it is violated. In our implementation, we generate invariants for every single passed source code line.

$$inv_{line}(angle) = -360 \leq angle \leq 360 \tag{4.10}$$

However, with few reference runs, it is difficult to build range invariants. Relation invariants between variable/value pairs check the relation between two numeric variables. Variable relations are often checked in conditional branches. This relation may be that a variable *angle* is always bigger than a variable *lazy* in a specific source line (4.11).

$$inv_{line}(angle) = lazy < angle \tag{4.11}$$

Our analysis compares each variable value to all other variable values to detect the relations between the variables. In our Navit example, the method *vehicle_update* is always called with the fourth variable *lazy* which defines the drawing mode and is, for most cases, 0 or 1. Comparing the variable *angle* against the variable *lazy*, shows that, before operation 2, $angle < lazy$. After operation 2, $lazy < angle$ is fulfilled for every nonfailure run. However, for the failing run, $angle < (lazy == 0)$ still holds. Figure 4.15 shows the sequence of variable relations between the variables *angle* and *lazy*.

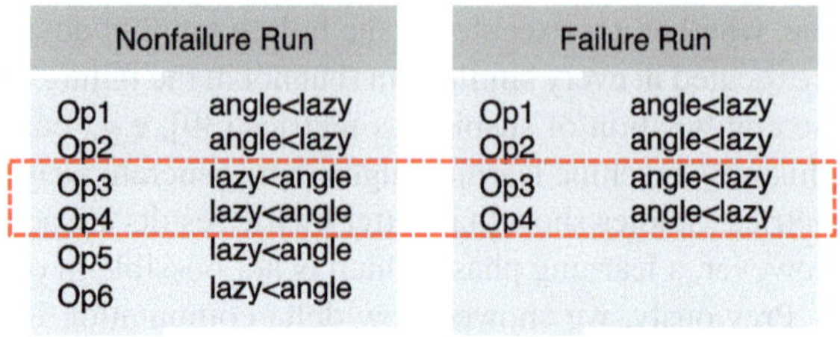

Fig. 4.15 Detecting invariant deltas between two Navit GPS processing

In Fig. 4.15 the wrong relations in lines 3 and 4 in the failure run point exactly to the root-cause of the failure. The resulting analysis report includes anomalies detected by comparing the coverage of the failing run to the similar runs. Additionally, the report includes the generated relation invariants which differ between the failing run and the similar runs.

4.5.2.5 Accelerated Monitoring

This section presents how to accelerate software monitoring based on [13, 43]. Many approaches use special hardware to monitor the embedded software under test. However, this hardware can be expensive or is even not available for new platforms. Therefore, we present an approach on how to optimize the monitoring to achieve fast dynamic verification results. Most developers already use an incremental approach to manually debug software. They first set breakpoints on methods starting to detect anomalies in the method call sequence. Afterwards, they examine suspect methods and they stepwise refine their examination. Our tool achieves accelerated monitoring for bug root-cause analyses, by applying this concept in an automated way. First, we define the basic concept for single-level (SL) monitoring.

> **Single-Level Monitoring—SL**: Monitoring single steps through every executed source code line during a (processing) run. It monitors the current variable values for every monitored source code line.

The analyses presented in the previous section can be applied on the traces generated by *SL* monitoring. However, running single-level monitoring can be slow. The methods of the software are usually executed much less frequently than the source code lines. Therefore, it is usually faster to monitor the methods instead of monitoring every source code line.

MultiLevel Monitoring—ML: In the first replay, the method calls are monitored. The following activities can be applied on this method calls trace: *Detect Failure Run* and *Detect Similar Runs*. In a second replay, the failure run as well as the similar runs are monitored in detail with single-step monitoring. The *Delta Computation* can be applied on this generated trace. This way, all runs despite the failure and the similar runs do not have to be monitored in detail.

The concept is illustrated in Fig. 4.16. It shows two replays: one for method level monitoring and one for monitoring the relevant runs in detail with single-stepping. On method level, first, the failure run is detected (A.) by comparing it to a nonfailure replay, as presented in Sect. 4.5.2.2. The failure run in Fig. 4.16 is *Run*7. Afterwards, the similar runs to the failing run are identified on method level (B.), as presented in Sect. 4.5.2.3. Here, the similar run is *Run*5. However, our approach can detect and handle several similar runs. The difference between the failing run and the similar runs (*Run*5 and *Run*7) is implemented by single-step monitoring (same as *SL*) and analyzing every source code line as presented in Sect. 4.5.2.4. Thus, in the example, *Run*6 and *Run*8 are not monitored in detail.

However, pausing at every passing of method causes a high monitoring time as well (like presented in Sect. 4.5.4), especially when short methods are called in high frequency. Thus, in the following, we propose a concept for debugger-based efficient monitoring of the method coverage of a run.

ML Method Monitoring—MLMethd: Method coverage monitoring traces every executed method in a run only once.

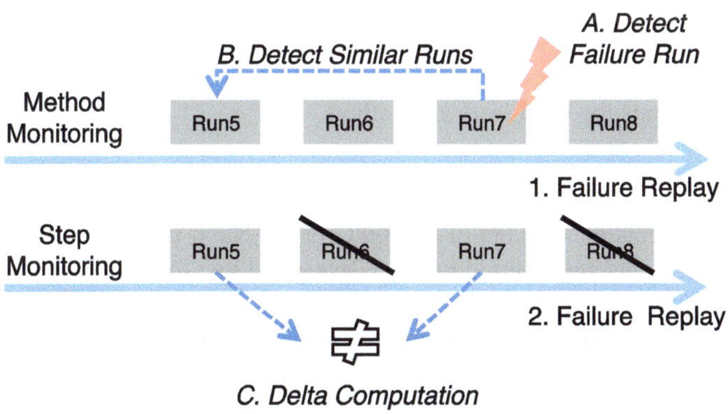

Fig. 4.16 Multi-level (ML) monitoring

Algorithm 1 shows the concept for *MLMethd* in pseudocode. This way, the monitoring of method coverage is efficient, because the monitoring tool only has to trace every method once.

Algorithm 1 Efficient method coverage monitoring for a run

Require: *allmethods* = List of all methods to monitor
Require: *context* = Monitoring context
Ensure: *methodcov* = Set of covered methods in the run
  $\forall_{m \in allmethods} context.setBreakpoint(m)$
  while *context.nextBreak*() **do**
    $where = context.where()$
    $methodcov = methodcov \cup where$
    $context.removeBreakpoint(where)$
  end while

However, after the failure run is detected and the similar runs are computed, the delta computation step requires a fine-grained trace. The monitoring of this trace can be very slow. A stepwise refinement can accelerate the monitoring. *MLMethd* results in a list of suspect methods (*relmethds*) which are either executed in the failing run or in the similar runs.

ML Backtrace Monitoring—MLBack: The backtraces of the methods in *relmethds* are identified. Every method which occurs in those backtraces is first monitored without stepping into the called methods (side steps). Methods which are executed either in the similar runs or the failing run are not monitored (no comparison is possible). *MLBack* includes *MLMethd*.

For our Navit example, we consider five different methods:

- *update*: Updates the current vehicle state based on GPS input data.
- *route*: Represents the routing calculation.
- *vehi.*: Represents the *vehicle_draw* method.
- *set*: Changes the drawing mode.
- *draw*: Invokes the draw of the vehicle pointer.

Figure 4.17 presents the concept of *MLBack* for the Navit bug. Every black line (or solid line) represents a monitored method. Every red line (or dotted line) represents a not monitored method.

First, a replay monitors the method coverage and detects that the *draw* method misses in the failing run. Several similar runs to the failing run are detected based on method monitoring. During *MLBack*, the methods which occurred in the backtrace of the invocation of *draw* in the nonfailure runs are collected in *relmethds*. During a second replay, the methods in *relmethds* are monitored in the similar runs and the failing run without stepping into the called (or side step) methods. During stepping

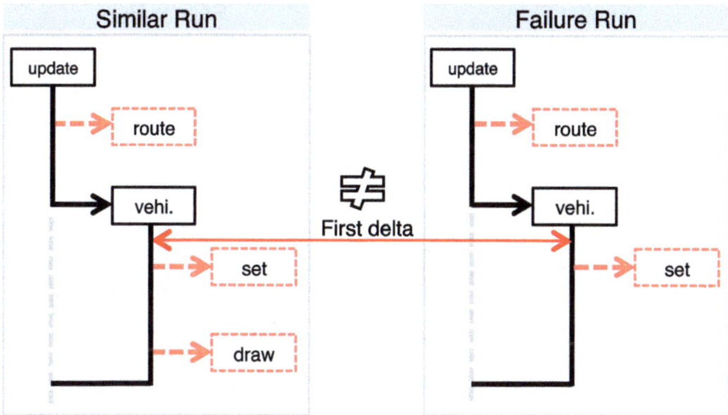

Fig. 4.17 Multi-level backtrace (MLBack) monitoring

through these methods, the values of local variables and method parameter variables are monitored. *MLBack* results in a list of methods which contain anomalies *suspectmethd*. The methods *update* and *vehi.* are monitored without stepping into side steps (here, *route* and *set*). This way, the routing calculation which is not relevant for our considered bug, but would require a lot of operations, is not monitored. Additionally, *draw* is not monitored, because it does only occur in the similar runs, but does not occur in the failure run (and cannot be compared). A first anomaly in the variable values can be detected after the calculation $angle+ = 360$ in *vehi.*

Definition ML Step Monitoring—MLStep: This monitoring concept has a list of suspect methods *suspectmethd* as input. These methods are monitored in an additional replay with the called side steps. Methods which are executed either in the similar runs or the failing run are still not monitored.

In the Navit example, a third replay is executed to additionally monitor the side steps in suspect methods (see Fig. 4.18).

Here, the method *vehi.* is additionally monitored with side steps. This includes the stepping into the method *set* (and some other methods not presented in Fig. 4.18). In an alternative implementation of Navit, the method *set* (or another method *even* which is called by *vehi.*) may include the wrong source code line with $angle+ = 360$, after which wrong relations between *angle* and *lazy* are detected.

4.5.3 Implementation

This section presents how a set of runs in a replay is monitored by single-stepping over every source code line (Single-Stepping Monitoring). On source code level,

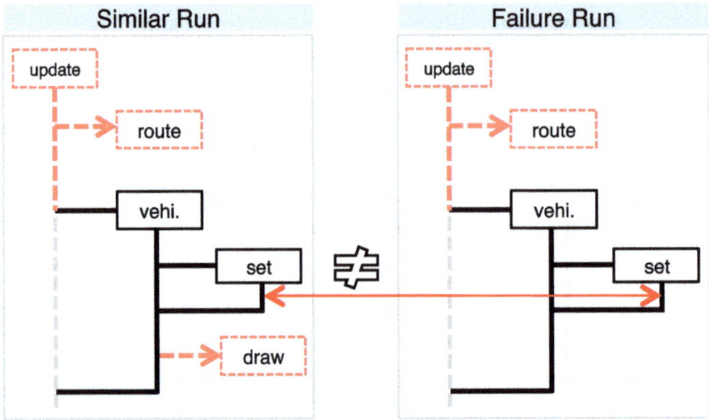

Fig. 4.18 Multi-level step (MLStep) monitoring

the variable values are monitored for every executed source code line in a run. This implementation is required for *SL* as well as for *ML* in the second replay. We show how method level monitoring (Method Coverage Monitoring) is implemented. On method level, the coverage of executed methods in a run is monitored. Additionally, we show how *MLBack* and *MLStep* can be implemented using GDB Python scripts.

4.5.3.1 Single-Stepping Monitoring

Listing 8 shows the implementation for single-stepping monitoring for the Navit example (considering the GPS processing). Lines 1–2 set breakpoints at the locations in the source code where the GPS processing starts and ends. In case the location where the processing starts is reached (line 7), the monitoring is activated (line 8). If the monitoring is activated, every step in the processing is monitored by printing local and argument variables (lines 13–14). The next source code line in the software under test is reached by executing the GDB *step* command (line 16).

4.5.3.2 Method Coverage Monitoring

Listing 9 shows the implementation for method coverage monitoring for the Navit example (considering the GPS processing). Lines 1–2 set breakpoints at the locations in the source code where the GPS processing starts and ends. Additionally, breakpoints are set on every method of the Navit software (lines 5–6); they are disabled at the beginning of the execution (line 7). If the location where the processing starts is reached (line 10), the monitoring is activated (line 11) and, additionally, the breakpoints for every method in the Navit software are enabled (line 12). If only the coverage of executed methods should be monitored, every breakpoint could be disabled after first passing (lines 18–19).

Listing 8 Implementation of single-stepping monitoring

```
gdb.execute("break␣gpsprocessingstart")
gdb.execute("break␣gpsprocessingend")
monitor=False
while running:
  where=gdb.execute("where",to_string=True)
  if "gpsprocessingstart" in where:
    monitor=True
    gdb.execute("step")
  elif "gpsprocessingend" in where:
    monitor=False
  elif monitor:
    locs=gdb.execute("info␣locals",to_string=True)
    args=gdb.execute("info␣args",to_string=True)
    trace.write(where,locs,args)
    gdb.execute("step")

...    # Sensor and Thread Replay Code
```

Listing 9 Implementation of method coverage monitoring

```
gdb.execute("break␣gpsprocessingstart")
gdb.execute("break␣gpsprocessingend")
monitor=False
for m in navitmethods:
   point=gdb.execute("break␣%s"%m)
   disable(point)
while running:
 where=gdb.execute("where",to_string=True)
 if "gpsprocessingstart" in where:
   monitor=True
   enableAllMethodBreakpoints()
 elif "gpsprocessingend" in where:
   monitor=False
   disableAllMethodBreakpoints()
 elif monitor:
   trace.write(where)
   if coverage:
     disable(where)
   gdb.execute("cont")

...    # Sensor and Thread Replay Code
```

4.5.3.3 Method Backtrace Monitoring

Listing 10 shows the implementation for method backtrace monitoring for the Navit example (considering the GPS processing). Lines 2–3 set breakpoints on all methods in *relmethds* reported from previous method monitoring and analysis. These methods are stepped through with the command *next* of the GDB (line 14), while monitoring variable values and without stepping into called methods (side steps). The implementation of *MLStep* is similar, but monitors the methods reported from the *MLBack* analysis. It uses the command *step* of the GDB instead of the command *next*.

Listing 10 Implementation of MLBack

```
1   ...
2   for m in relmethd:
3      point=gdb.execute("break␣%s"%m)
4      disable(point)
5   while running:
6     where=gdb.execute("where",to_string=True)
7     if "gpsprocessingstart" in where:
8       monitor=True
9       for b in breaks:
10        gdb.execute("enable␣%s"%b)
        ...
11    elif monitor:
12      locs=gdb.execute("info␣locals")
        ...
13      trace.write(where,locs,args)
14      gdb.execute("next")

...   # Sensor and Thread Replay Code
```

4.5.4 *Experiments*

The previous sections showed how dynamic verification supports the developer to analyze the root-causes of bugs. In Sect. 4.5.2.5, we presented optimization techniques to accelerate the monitoring for those analyses. We tested the multi-level monitoring (ML) concept for the Replace tool of the Siemens Test Suite [22]. The Replace program is delivered in 32 different versions and with several test cases. We tested 19 of the first 20 versions, which all contain one bug (we could not manually detect a bug in version 19). To generate a possible random replay, we implemented a replay generator which randomly selects 99 test cases from the 5542 test cases in the Siemens Test Suite fault matrix which do not cause the failure. In the end of the replay, we added as run 100 a failing run which executes the bug. For the random selection of test cases, we used the Python random function with seed 100 for every generated replay. We generated a second replay with 200 nonfailing runs to simulate

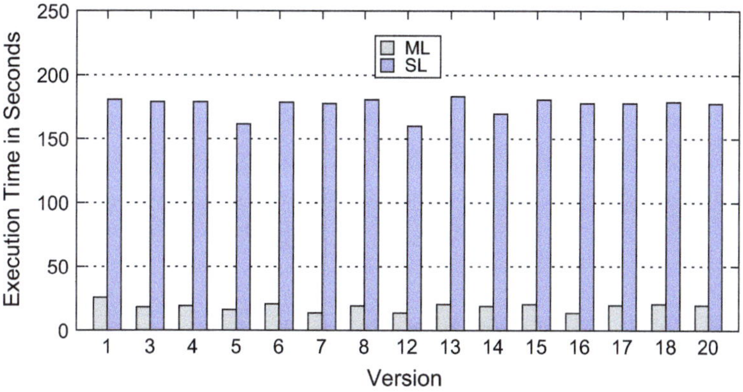

Fig. 4.19 Monitoring runtime for multi-level analyses for replace

noise (using same seed 100). We used this replay as nonfailure replay for the failing run detection presented in Sect. 4.5.2.2. For SL, we monitored the execution of the 100 replay runs collecting source code line coverage as well as numeric local and argument variable values. For ML, the failing run was detected based on method coverage. For this implementation, we additionally used the count of occurrences of a specific method for building hamming words (as presented in Sect. 4.5.2.3 using *threshold* $= 5$). The three most similar runs to the failure run were determined based on method coverage as well. In a second replay, the failing run and the similar ones were monitored and compared considering line coverage and variable relations. We observed that ML monitoring is much faster than SL monitoring. Figure 4.19 shows the runtime for single-level (SL) monitoring compared to multi-level (ML) monitoring for Replace (for those versions where we could detect the respective bug, using our algorithms).

For a general evaluation, only the monitoring runtimes were included, because the runtimes for analyzing the monitoring result depend on the type of analysis. ML consists of a first replay monitoring on method level and a second replay monitoring on line level. ML and SL monitored all local variables, arguments, and macros for the delta computation for the failing run and the three most similar runs. We found that ML monitoring is much faster than SL monitoring. For Replace, a monitoring acceleration of ∅ 9.5X was achieved. Table 4.8 shows the number of reported suspect lines of SL and ML and the hamming distance from the failing run to the three similar runs for the different Replace versions.

δ-ML/SL presents the count of reported suspect lines. *Fail Detect* shows whether the failing run could be detected on method level. *Dist. Meth.* and *Dist. Line* show the hamming distances to the computed similar runs. *Detect* presents whether the buggy lines are detected with the coverage delta computation or the variable relation delta computation. In six versions (3, 4, 7, 13, 14, 16), the detected similar runs were the same for ML and SL. In these cases, the reported suspect lines were the same. In the 15 experiments, every bug of the version was detected with SL and ML, either pointing to the coverage delta (Cov) or to a delta in variable relations (Rel). Four reports for bugs indirectly pointed to the failure (Version 12—many occurrences of MAXPAT

Table 4.8 Reported delta lines for replace and hamming distances

Example	V1	V3	V4	V5	V6	V7	V8	V12	V13	V14	V15	V16	V17	V18	V20
δ-ML	89	109	117	34	83	71	50	102	119	119	62	71	47	105	47
Fail detect	✓	✓	✓	✓	✗	✓	✓	✓	✓	✗	✓	✓	✓	✗	✓
Dist. Meth.	0,0,1	0,0,0	1,1,1	0,1,1	0,0,0	5,5,5	2,2,2	6,6,6	0,0,0	0,0,0	3,3,3	5,5,5	0,0,0	0,0,0	0,0,0
Dist. line	18,36,36	13,11,11	22,24,24	44,22,11	19,37,11	48,51,51	39,15,45	53,51,66	15,17,17	9,21,17	36,46,43	48,51,51	26,2,9	16,9,21	26,2,9
δ-SL	118	109	117	66	73	71	53	97	119	119	41	71	51	131	51
Dist. line	18,19,23	11,11,13	22,24,24	11,22,22	9,11,11	48,51,51	15,15,15	49,51,53	15,17,17	9,17,21	36,36,38	48,51,51	2,2,5	9,10,12	2,2,5
Detect	Cov	Rel	Rel	Rel	Cov	Rel	Rel	iRel	Cov	iRel	iRel	Rel	Rel	iRel	Rel

in relation deltas, Versions 15 and 18—relation difference of other variables in buggy line, Version 14—difference of relations in enclosed line of if-clause). We tested our tooling with the four other versions of Replace (2, 9, 10, 11), but in these tests, the buggy source code line could not be detected with our delta analysis with ML and SL. Here, the bugs are mainly caused by predicates in if-clauses which access arrays (which are only available in registers). The row *Fail Detect* shows the versions, for which the failing run could be detected on method level. The failing run could not be detected for the versions 6, 14, and 18 with the optimized lightweight classification presented in Sect. 4.5.2.2 based on a randomized selected nonfailure replay. Thus, 63% of the bugs could be automatically and efficiently localized with ML based on one failure replay with 100 runs and one nonfailure replay with 200 runs.

In some cases, delta computation for ML reported less false positives, because the similar runs detected by SL coincidentally caused more variable relation differences. In general, the report quality difference between SL and ML mainly depends on the composition of different runs in the replay. Note: Delta computation (cov+rel) for SL between the failing and every run in the replay would cause higher runtime for parsing and comparison of monitoring results (delta computation runtime without monitoring of three runs was ∅ 13.7 s for Replace).

However, the ML monitoring can still be too slow for monitoring instruction-intensive calculations. For example, the monitoring of a routing calculation of Navit on method level (which calls transformation methods in high frequency), can be very slow when the execution pauses at every method call. We measured the method monitoring for 20 GPS coordinates processing including the routing calculation, which required more than 50 h on an NVIDIA Tegra K1 ARM platform.

Figure 4.20 shows our experiments for the acceleration of the monitoring for root-cause analyses using refinement for the Navit software (*MLBack* and *MLStep*).

We measured a replay with 20 GPS and one with 30 GPS coordinates, both containing the Navit bug from the previous sections. In these experiments, each analysis (*MLBack*, *MLStep* and *SL*) detected the root-cause of the bug. These measurements include the routing calculation of the Navit GPS processing.

MLBack is 1354X faster than *SL* for the 20 GPS scenario. *MLStep* is 334X faster than *SL* for the 20 GPS scenario. For the 30 GPS scenario, the acceleration factors

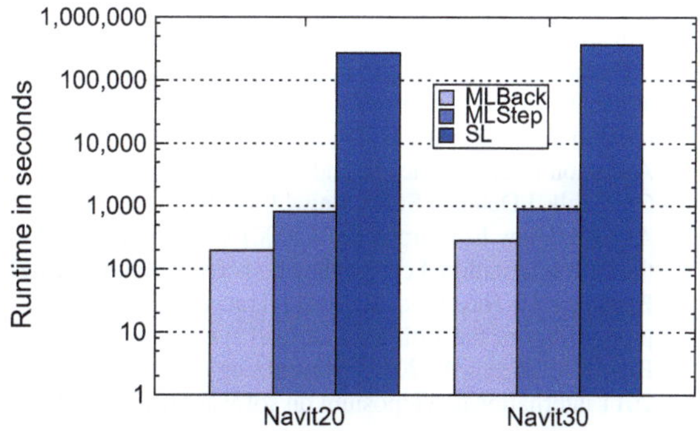

Fig. 4.20 Acceleration of monitoring using refinement for root-cause analyses for the Navit software [13]

are: *MLBack* 1292X and *MLStep* 405X. The acceleration does not increase with the longer sequence, because some GPS processing at the beginning take longer (for recalculating and redrawing of the routing line). The Navit measurements show that analyses with high monitoring overhead can be accelerated with hierarchical refinement, resulting in practicable dynamic verification performance.

4.6 Summary

The approach presented in this chapter showed how the manual debugging process can be supported by automated tools. We presented state-of-the-art approaches for automated bug reproduction. However, these approaches are mainly developed for specific platforms. Therefore, we developed the debugger-based approach which is portable to different embedded platforms. It reproduces sensor inputs and implements randomized thread scheduling for efficient concurrency bug localization. The chapter described how assertions can be implemented with a debugger tool to locate the cause of reproduced concurrency bugs. Afterwards, we showed how the root-cause of bugs can be located without needing specifications or assertions. The root-cause can be tracked down to changes in variable values which cause the bug. Based on a navigation software, we demonstrated how these root-cause localization techniques can be accelerated. This way, the application of slow monitoring tools can be optimized to make them applicable in practice.

The presented approach requires little adaptations for other software. However, the implemented scripts are all very short (every record/replay or monitoring script is shorter than 200 LOC). Thus, the tooling is highly extendable. Additionally, the GDB is supported by most embedded platforms and our script implementations are applicable on most embedded platforms. With few modifications, we could run all our scripts on an ARM Linux the same way as on an X86 Linux.

References

1. Navit-car navigation system. http://www.navit-project.org. Accessed Aug 2016
2. Abreu R (2009) Spectrum-based fault localization in embedded software. PhD thesis, University Delft
3. Amiar A, Delahaye M, Falcone Y, du Bousquet L (2013) Fault localization in embedded software based on a single cyclic trace. In: ISSRE '13: proceedings of the 24th international automated reproduction and analysis of bugs in embedded software 39 symposium on software reliability engineering. IEEE, pp 148–157
4. Anderson P (2008) The use and limitations of static-analysis tools to improve software quality. CrossTalk J Defence Softw Eng 42(4):18–21
5. Anvik J, Hiew L, Murphy GC (2006) Who should fix this bug? In: ICSE '06: proceedings of the 28th international conference on software engineering. ACM, pp 361–370
6. Barringer H, Havelund K (2011) Tracecontract: a scala dsl for trace analysis. In: FM '11: proceedings of the 17th international symposium on formal methods. Springer, pp 57–72
7. Burger M, Zeller A (2011) Minimizing reproduction of software failures. In: Proceedings of 2011 international symposium on software testing and analysis, pp 221–231

8. Charette RN (2009) This car runs on code. IEEE Spectr 21(6)
9. Dallmeier V, Lindig C, Zeller A (2005) Lightweight bug localization with ample. In: AADEBUG '05: proceedings of the sixth international symposium on automated analysis-driven debugging. ACM, pp 99–104
10. Dovgalyuk P (2012) Deterministic replay of system's execution with multi-target qemu simulator for dynamic analysis and reverse debugging. In: CSMR '12: proceedings of the 16th European conference on software maintenance and reengineering. IEEE, pp 553–556
11. Ebert C, Jones C (2009) Embedded software: facts, figures and future. Computer 42(4):42–52
12. Eichelberger H, Kropf T, Greiner T, Rosenstiel W (2013) Runtime verification driven debugging of replayed errors. In: ICTSS '13: proceedings of the PhD workshop of ICTSS' 13
13. Eichelberger H, Kropf T, Ruf J, Greiner T, Rosenstiel W (2015) Efficient fault localization during replay of embedded software. In: SEAA '15: proceedings of the 41th euromicro conference series on software engineering and advanced applications. IEEE, pp 43–52
14. Eichelberger H, Ruf J, Kropf T, Greiner T, Rosenstiel W (2014) Debugger-based record replay and dynamic analysis for in-vehicle infotainment. In: ICCSA '14: Proceedings of the 14th international conference on computational science and its applications. Springer, pp 387–401
15. Foundation G (2016) Gdb: the gnu project debugger. http://www.sourceware.org/gdb. Accessed Aug 2016
16. Foundation G (2016) Gnu pth—the gnu portable threads. http://www.gnu.org/software/pth/. Accessed Aug 2016
17. Goeders J, Wilton S (2014) Effective fpga debug for high-level synthesis generated circuits. In: FPL '14: proceedings of the 24th international conference on field programmable logic and applications. IEEE, pp 1–8
18. Goll J (2012) Methoden des software engineering. Springer, Wiesbaden
19. Gomez L, Neamtiu I, Azim T, Millstein T (2013) Reran: timing- and touch-sensitive record and replay for android. In: ICSE '13: proceedings of the 35th international conference on software engineering. ACM, pp 72–81
20. Heckeler P, Eichelberger H, Schlich B, Kropf T, Ruf J, Huster S, Burg S, Rosenstiel W (2013) Accelerated model-based robustness testing of state machine implementations. ACM Appl Comput Rev 13(03):50–67
21. Hower D, Hill M (2008) Rerun: exploiting episodes for lightweight memory race recording. In: ISCA '08: proceedings of the 35th international symposium on computer architecture. ACM/IEEE, pp 265–276
22. Hutchins M, Foster H, Goradia T, Ostrand T (1994) Experiments of the effectiveness of dataflow- and controlflow-based test adequacy criteria. In: ICSE '94: proceedings of the 16th international conference on software engineering. IEEE, pp 191–200
23. Jiang B, Long X, Gao X, Liu Z, Chan W (2011) Floma: statistical fault localization for mobile embedded system. In: ICACC '11: proceedings of the 3rd international conference on advanced computer control. IEEE, pp 396–400
24. Jones C (2012) A short history of the cost per defect metric. http://www.ifpug.org/Documents/Jones-CostPerDefectMetricVersion4.pdf. Accessed Aug 2016
25. Jones JA, Harrold MJ, Stasko J (2002) Visualization of test information to assist fault localization. In: Proceedings of 2002 international conference on software engineering, pp 467–477
26. Joorabchi ME, Mirzaaghaei M, Mesbah A (2014) Works for me! characterizing non- reproducible bug reports. In: MSR '14: proceedings of the 11th working conference on mining software repositories. IEEE, pp 62–71
27. Laadan O, Viennot N, Nieh J (2010) Transparent, lightweight application execution replay on commodity multiprocessor operating systems. In: SIGMETRICS '10: proceedings of the 2010 ACM SIGMETRICS international conference on measurement and modeling of computer systems. ACM, pp 155–166
28. Lee YH, Song YW (2010) Replay debugging for multi-threaded embedded software. In: EUC '10: proceedings of the 2010 IEEE international conference on embedded and ubiquitous computing. IEEE, pp 15–22

29. Leucker M, Schallhart C (2009) A brief account of runtime verification. J Logic Algebraic Program 78(5):293–303
30. Liggesmeyer P (2009) Software-qualitaet. Spektrum Akademischer Verlag, Heidelberg
31. Liu C, Yan X, Yu H, Han J, Yu P (2005) Mining behavior graphs for "backtrace" of noncrashing bugs. In: SDM '05: proceedings of the 2005 SIAM international conference on data mining
32. Liu X, Lin W, Pan A, Zhang Z (2007) Wids checker. In: Proceedings of 4th USENIX conference on networked systems design and implementation, pp 257–270
33. Maeng J, Kwon JI, Sin MK, Ryu M (2009) Rt-replayer: a record-replay architecture for embedded real-time software debugging. In: SAC '09: proceedings of the 2009 ACM symposium on applied computing. ACM, pp 1670–1675
34. Patil H, Pereira C, Stallcup M, Lueck G, Cownie J (2010) Pinplay: a framework for deterministic replay and reproducible analysis of parallel programs. In: CGO '10: proceedings of the 8th international symposium on code generation and optimization. IEEE/ACM, pp 2–11
35. Sen K, Kalasapur S, Brutch T, Gibbs S (2013) Jalangi: a selective record-replay and dynamic analysis framework for javascript. In: ESEC/FSE '13: proceedings of the 9th joint meeting on foundations of software engineering. ACM, pp 488–498
36. Shin H, Endoh Y, Kataoka Y (2007) Arve: aspect-oriented runtime verification environment. In: Proceedings of 2007 runtime verification, pp 87–96
37. Tan L, Liu C, Li Z, Wang X, Zhou Y, Zhai C (2014) Bug characteristics in open source software. Imperical Softw Eng 19(6):1665–1705
38. Wu J, Geyer C, Rehg JM (2011) Real-time human detection using contour cues. In: ICRA '11: proceedings of the 2011 international conference on robotics and automation. IEEE, pp 860–867
39. Xuan J, Monperrus M (2014) Learning to combine multiple ranking metrics for fault localization. In: ICSME '14: proceedings of the 30th international conference on software maintenance and evolution. IEEE, pp 191–200
40. Yasushi S (2005) Jockey: a user-space library for record-replay debugging. In: AADEBUG '05: proceedings of the sixth international symposium on automated analysis-driven debugging. ACM, pp 69–76
41. Zeller A (2009) Why programs fail: a guide to systematic debugging, 2nd edn. Morgan Kaufmann Publishers
42. Zhang S, Ernst MD (2013) Automated diagnosis of software configuration errors. In: Proceedings of 2013 international conference on software engineering, pp 312–321
43. Zuo Z, Khoo SC, Sun C (2014) Efficient predicated bug signature mining via hierarchical instrumentation. In: ISSTA '14: proceedings of the 2014 international symposium on software testing and analysis. ACM, pp 215–224

Chapter 5
Model-Based Debugging of Embedded Software Systems

Padma Iyenghar, Elke Pulvermueller, Clemens Westerkamp, Juergen Wuebbelmann and Michael Uelschen

5.1 Introduction

Software engineering has gone through several paradigm shifts (assembly language → structural programming → object-oriented → model-driven development (MDD)), driven by the requirements for building more complex systems. The inherent necessity to achieve reliable systems, inline with these paradigm shifts, has ushered in copious verification and validation techniques.

However, debugging and runtime monitoring remains the widely used process for finding and resolving defects (bugs) that prevent correct operation of the underlying software system. Debugging tools, in general, help to identify errors at the various stages of the software development process. Some of the commonly used traditional debugging tools involve "printf" statements, data monitors, and operating system monitors [17]. On the other hand, some sophisticated techniques available in the embedded software development tools are profilers, memory testers, and execution tracers [5, 8, 32], to mention a few.

P. Iyenghar (✉) · E. Pulvermueller
Software Engineering Research Group, University of Osnabrueck, 4469, 49069 Osnabrueck, Germany
e-mail: piyengha@uos.de

E. Pulvermueller
e-mail: elke.pulvermueller@uos.de

C. Westerkamp · J. Wuebbelmann · M. Uelschen
University of Applied Sciences, 1940, 49009 Osnabrueck, Germany
e-mail: c.westerkamp@hs-osnabrueck.de

J. Wuebbelmann
e-mail: j.wuebbelmann@hs-osnabrueck.de

M. Uelschen
e-mail: m.uelschen@hs-osnabrueck.de

D. Lettnin and M. Winterholer (eds.), *Embedded Software Verification and Debugging*, Embedded Systems, DOI 10.1007/978-1-4614-2266-2_5

Observing and examining the behavior of software execution at runtime can be termed as runtime or online software monitoring and verification. Embedded systems present particular challenges for monitoring, which necessitate a nonintrusive or a minimally intrusive monitoring methodology. This is especially true for resource constrained, deeply embedded systems (e.g., 16-bit systems with 64 KiByte memory). Runtime monitoring techniques used for several application purposes [5, 8, 32] often employ a target output/computer to diagnose and interpret the results in formats such as plain text and graphs. However, with the advent of model-driven methodologies, the applicability and usage of models, such as Unified Modeling Language (UML) [4] diagrams, are under evaluation for model-based runtime monitoring and debugging of software systems.

In the field of real-time embedded software systems, model-based debugging and visualizing embedded software systems, using diagrams, such as sequence and timing diagrams, presents an exciting outlook. Model-based visualization of embedded software system behavior (at runtime), using sequence diagrams and state charts, is possible in proprietary MDD tools, such as Rhapsody [14]. However, such existing runtime monitoring and debugging methodologies are not well-suited for applicability in deeply embedded systems, primarily because of the monitoring overhead involved. The following section enumerates the drawbacks of debugging, runtime monitoring, and visualization of target behavior in real time, in the state-of-the-art tools and methodologies.

5.1.1 Problem Statement

The drawbacks of model-based debugging and runtime monitoring, which involve extensive source code instrumentation in the underlying software, are discussed below. The crux lies in the applicability of such techniques in deeply embedded software systems. An example of such a resource constrained embedded system is a 16-bit system with less than 64 KiByte memory.

- **Significant instrumented code size** The instrumented code (for debugging and monitoring) increases with an increase in the application size. The instrumented code varies based on the application size and complexity. Hence, there arises a question of scalability and applicability of such an approach in debugging small embedded software systems. For instance, an existing MDD tool [14], while supporting model-based monitoring of embedded systems, makes use of techniques such as dynamic source code instrumentation or downloading a significant instrumented code on the target.
- **Requirement for sophisticated interfaces**: Often, sophisticated interfaces and communication protocols are required to download the instrumented code and visualize the behavior of the target in real time at the host computer. For instance, let us consider the MDD tool Rhapsody [14], which provides a “live-animation” feature for model-based visualization of the target behavior in real time.

It introduces significant (and dynamic) source code instrumentation overhead. Further, it requires sophisticated debug communication interfaces (e.g., TCP/IP over Ethernet) to download the debug code on the target and visualize the behavior at the host, using UML diagrams. It is intuitive to perceive that such techniques would further result in protocol and performance overhead during debugging and/or runtime monitoring. Moreover, such interfaces are not necessarily available in memory-size constrained, deeply embedded targets.
- **What you verify is not what you deliver**: The instrumented code is not only significant, it is often removed after the debugging process/verification is completed. This implies that the behavior of the RTESS during debugging may not remain the same after the debugging process is complete. For instance, there could be a change in the program-instructions or clock cycles before which the specific system code is executed (because of the execution cycles of the instrumented code). This necessarily means that the system that is debugged/verified is not the same system that is delivered as the end-product.

Thus, the existing tools introduce unbounded overhead and highly intrusive monitoring mechanisms for model-based debugging and visualization of real-time behavior of targets, using UML diagrams. Clearly, such approaches are not suitable for applicability in deeply embedded systems.

5.1.2 Contribution

Based on the problems stated above, it is clear that there is a need for an integrated model-based debugging framework and monitoring approach, which provides scalable model-based debugging for deeply embedded systems. Such a monitoring methodology is referred to in this paper as the time and memory-size aware runtime monitoring.

One such model-based debugging approach, which addresses the aforementioned limitations, is discussed in [20]. With this model-based debugging approach, the behavior of memory-size constrained RTESS can be visualized in real time, using UML sequence and timing diagrams (at the design level using a minimally intrusive target monitor). Performance metrics and evaluation of the debugging approach proposed in [20] is discussed in [19]. This book chapter elaborates on the debugging approach presented in [19, 20] and extends it with the following novel contributions.

- A brief outline of the model-based debugging approach is provided.
- The requirements of a time and memory-aware runtime monitoring methodology, toward applicability for model-based visualization of target behavior, are elaborated.
- Two variants of the proposed monitoring methodology, i.e., (a) software and (b) on-chip monitoring are presented and their prototype implementation is discussed.

- Frame formats for sending the trace data between host and target in the proposed model-based debugging approach is elaborated.
- An experimental evaluation of the proposed monitoring mechanisms is provided. A discussion and evaluation of the proposed approach, in comparison with the existing approaches, is presented.

The remainder of this paper is organized as follows. Section 5.2 deals with related work. The model-based debugging methodology is outlined in Sect. 5.3. The runtime monitoring methodologies are discussed in Sect. 5.4. A prototype of the monitoring methodologies and an experimental evaluation are discussed in Sect. 5.5. The performance metrics of the proposed approach are discussed in Sect. 5.6. A discussion on the salient features of the proposed approach is presented in Sect. 5.7. Section 5.8 concludes this paper.

5.2 Related Work

The exponential growth of the embedded software systems necessitates the use of advanced and automated methods of development and testing. In this context, Model-Driven Architecture (MDA), proposed by the OMG [28], promises several advantages and superiority, superseding the traditional way of developing the embedded systems. This is supported by the studies conducted in [3, 9, 22, 23].

The MDA model is related to multiple standards including the UML. UML comprises of general purpose diagrams and profiles. UML profiles introduced by the OMG consist of a set of new stereotypes for a particular domain. The widespread applicability of UML (general purpose diagrams and profiles) as a modeling language for embedded systems is evident from the numerous studies in the literature [20, 22]. In our proposed model-based debugging methodology we use UML to specify the design model (e.g., class diagram, state charts, etc.). UML interaction diagrams, such as timing diagram and sequence diagram, are used for visualizing the target behavior in real time at the design level in our approach.

Irrespective of the evolution in embedded software engineering, the model-based tools for embedded systems continue to use monitoring approaches for applications such as debugging and testing. Software monitoring has been in use for over 35 years for a variety of domains and application purposes [5, 29]. Some of the domains in which runtime monitoring is applied include, distributed systems, fault-tolerant systems, real-time critical systems, and embedded systems [32].

Domains such as the embedded systems present particular challenges for monitoring, as the system internals may not be easily observable and have limited resources or real-time constraints. Hence, utmost care must be taken to avoid incurring extensive runtime overhead in the form of additional resources (e.g., memory, time). A technique for time-aware instrumentation of embedded software is discussed in [8]. It demonstrates how instrumentation can be used to maximize trace reliability and computing the minimal trace buffer size. However, most of the aforementioned monitoring approaches, except [8], concentrate on applications running on desktop computers. They do not consider their impact on the time or memory requirement of the applications.

Monitoring systems are classified according to the probes/instrumentation used, as hardware, software, hybrid, and on-chip monitoring. This classification is used both in early and late surveys on monitoring [32]. A time-aware, software-based instrumentation methodology is presented in [8]. Similarly, a software-based monitoring approach discussed in [20] is used for visualizing the behavior of targets in real time, using UML diagrams. Software monitoring with controllable overhead is discussed in [13]. A hybrid monitoring approach using a proprietary tool and In-Circuit Emulator (ICE) for testing of embedded systems software against UML models is discussed in [11]. A survey on monitoring approaches is presented in [32]. However, the applicability of the monitoring approach for visualizing target behavior in real time (esp. in deeply embedded targets) is missing in [8, 11, 13, 20, 32].

In this paper, a model-based debugging mechanism, which makes use of a target monitor in the embedded system and a target debugger at the host computer (design level), is discussed. Two variants of the target monitor, namely (a) software and (b) on-chip monitoring mechanisms, are presented. The target monitor sends trace data pertinent to the behavior of the target to the host side by back annotation. Some related work pertaining to target monitor and back annotation of trace data from target to host are discussed here. A study conducted in [12] deals with back annotations and continuous feedback about target behavior to the host side. This study is conducted on Java-based microprocessors for worst-case execution time (WCET) analysis. However, Java-based microprocessors are not necessarily the preferred choice in a memory-size constrained RTESS. In [21], an implementation of an event-driven hardware/software collaborative monitor system, enabling system-level monitoring on target at different abstraction levels is presented. In this work, the monitor system is claimed to collaborate seamlessly with other components in a model-driven testing tool chain. Though [12, 21] deal with back annotation and monitor systems respectively, a collaborative approach toward model-based debugging using a model-based target debugger is unavailable. Similarly, the suitability of runtime verification and monitoring approaches for embedded systems is discussed in [32]. Nevertheless, monitoring approaches for supporting model-based runtime visualization of embedded system behavior is missing in [32].

Commercial MDD-based tools such as [2, 7, 14] are limited in terms of debugging in real time at the design level for RTESS. Moreover, these tools and their model-based debugging feature cannot be used for small RTESS, because of memory, performance and protocol overhead (TCP/IP over Ethernet, etc.) with the target system. All these result in potentially inefficient communication between the target and the host. The dynamic source code instrumentation introduced by these tools could also result in affecting the real-time behavior of the RTESS.

Thus, it is evident that even though model-based development and debugging is being used for RTESS, applicability of model-based debugging approaches for deeply embedded systems is still in fledgling stages (both in academia and commercial tools).

5.3 Model-Based Debugging Framework

The proposed framework for model-based, design level debugging of deeply embedded systems discussed in this paper, is shown in Fig. 5.1. An outline of the proposed framework is provided here to place in context the main focus of this paper. This paper concentrates on a time and memory-aware runtime monitoring methodology for debugging and visualizing the (deeply embedded) target behavior in real time (on the host).

5.3.1 Overview

The proposed framework comprises of a target debugger on the host side with a runtime monitoring solution on the target side as seen in Fig. 5.1.

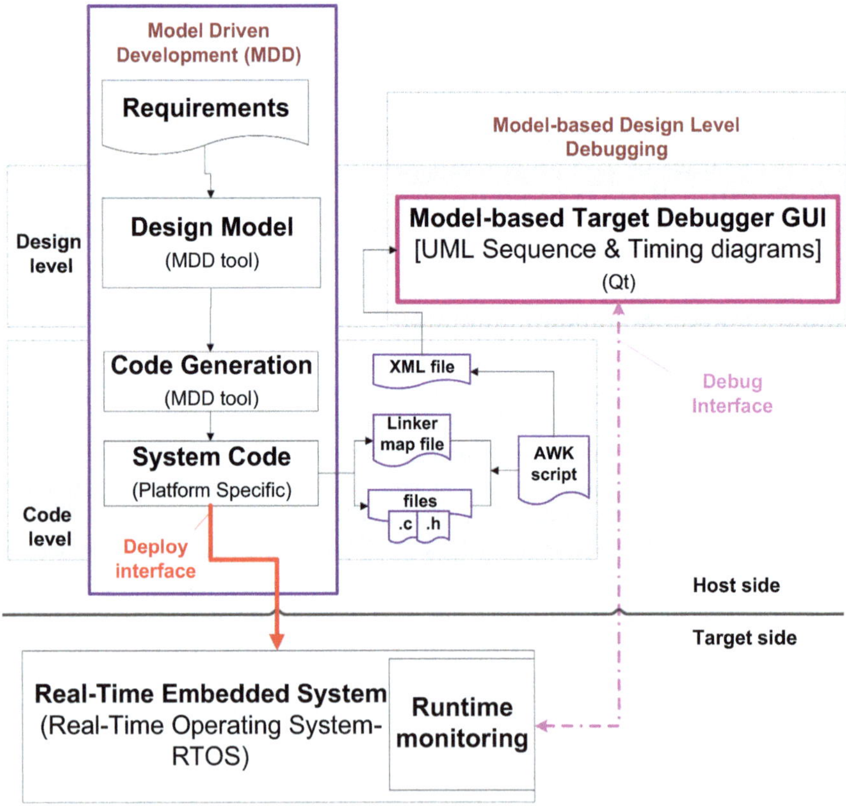

Fig. 5.1 Model-based design level debugging for embedded systems

5.3.1.1 MDD Phase and System Code

Based on the requirements, the design model for the embedded software application to be developed is specified (Fig. 5.1) using a modeling language (e.g., UML using a MDD tool Rhapsody [14]). The main functionality of the application can be specified, for example, in UML class diagrams. The detailed functionality and reactive behavior of each class can be represented as a state diagram, for example, using UML state charts. The next step is the automatic code generation process, which is most often supported in MDD tools (e.g., [14]). In the prototype, the system code is generated in C. The system code thus obtained can be executed on the target after cross-compilation. During the compilation of the system code, an XML file is generated at the host side (Fig. 5.1) using an AWK script (which parses the linker map file, source, and header files for a given project). The generated code is deployed using a deploy interface (such as the Keil Ulink [27] adapter) on the target, which runs a real-time operating system (RTOS). An RTOS framework suitable for resource constrained embedded systems, namely OORTX-RXF [33], is used in the prototype. However, this approach can also be applied to other UML-based modeling tools (e.g., [7]) and modeling alternatives, such as Matlab/Simulink [26] and LabView [24]. In the prototype, the system code is generated in the programming language C. Note that this approach is independent of the modeling language, the language in which the system code is generated and the underlying RTOS in the embedded system.

5.3.1.2 Target Debugger Graphical User Interface (GUI)

The model-based, design level target debugger graphical user interface (GUI) on the host side receives back-annotated trace data from the target monitor using a debug communication interface. The trace data provides details about the target behavior in real time. The target debugger reconstructs the behavior of the target in real time using UML interaction diagrams, such as the sequence diagram and the timing diagram in the GUI. The target debugger is implemented in the User Interface (UI) framework Qt [30].

The target debugger GUI consists of three blocks as shown in Fig. 5.2. Block (a) shows the classes, objects, states, and attributes available in the embedded software running on the target system. Block (b) displays the sequence of events and the temporal behavior of the target using UML sequence diagrams with time stamps (sequence diagram tab) and UML timing diagrams (timing diagram tab). Block (c) displays the reconstructed messages on the host side (based on back-annotated data from target and the XML file).

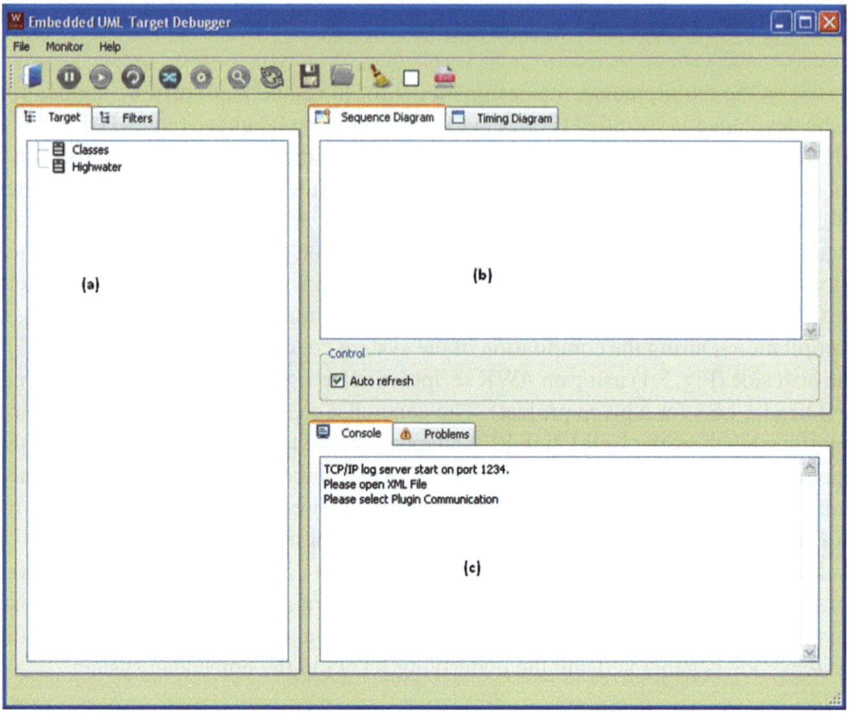

Fig. 5.2 Model-based target debugger GUI

5.3.1.3 Target Monitor

The target monitor is primarily used for runtime monitoring, i.e., to send pertinent trace data to the host about target behavior in real time. Two variants of the runtime monitoring methodology, (a) software and (b) on-chip monitoring are discussed in detail in Sect. 5.4.

5.3.1.4 Visualizing Target Behavior in Real Time

On the host computer, the animation program in the target debugger GUI is started. The XML file generated for the corresponding project is loaded in the target debugger. A debug communication interface (for the communication between target and host) is chosen. The target debugger is now ready for receiving the trace data from the target.

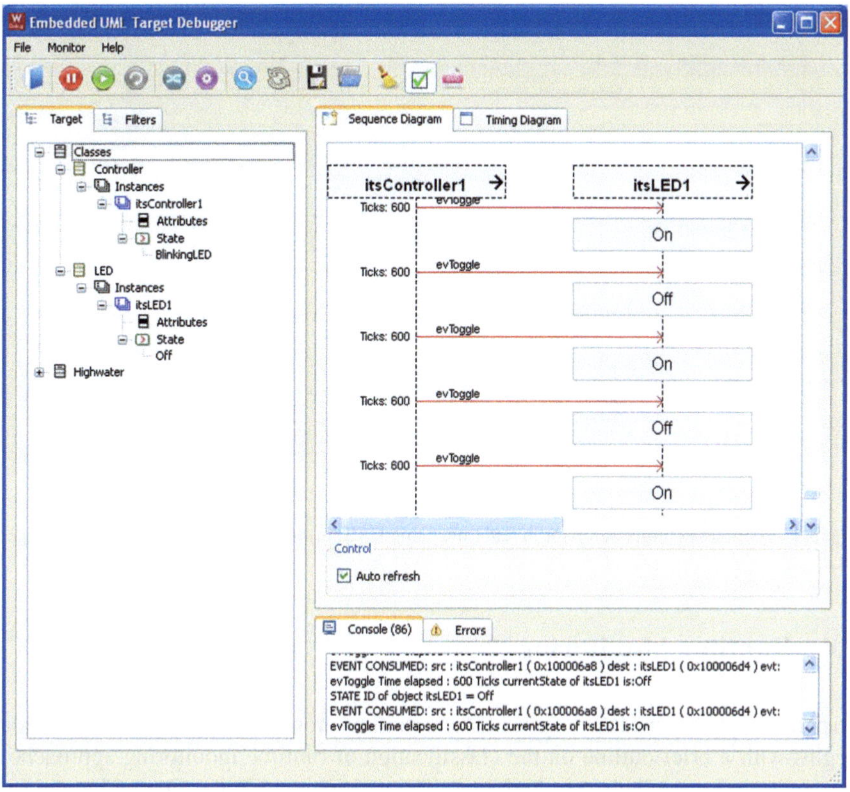

Fig. 5.3 Visualizing target behavior in real time using UML sequence diagrams in the target debugger GUI

Once the system code is deployed in the target, the target monitor (bundled with the RTOS framework, e.g., RXF) starts sending state, event and temporal information pertinent to the behavior of the target. This trace data is sent via the chosen debug communication interface to the host computer.

The animation program in the target debugger receives and decodes this trace information with the aid of the XML file. It reconstructs the target behavior on the host computer, at the design level, using UML sequence diagrams with time stamps and timing diagrams (Figs. 5.3 and 5.4). Thus, with the aid of this approach the target behavior can be visualized and debugged in real time at the host computer.

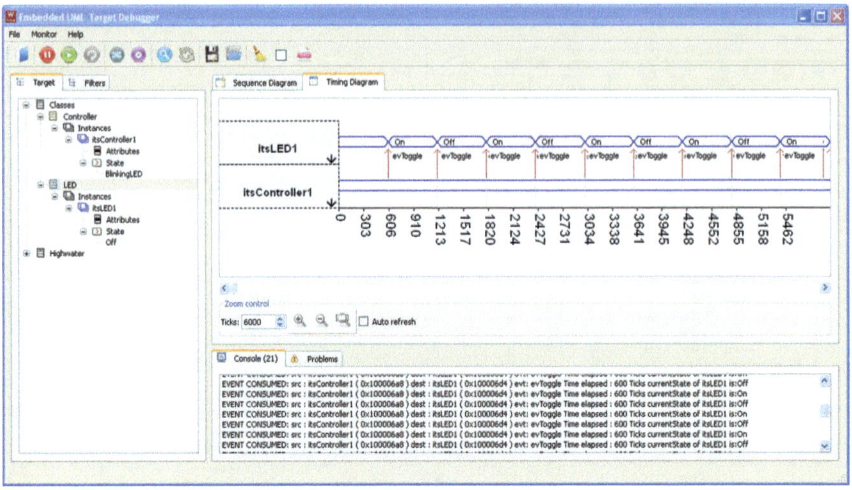

Fig. 5.4 Visualizing target behavior in real time using UML timing diagrams in the target debugger GUI

5.4 Runtime Monitoring

To perceive a time- and memory size-aware runtime monitoring approach, this section begins with a brief outline on the classification of runtime monitoring approaches and a discussion on their pros and cons. The requirements for a time-and memory-aware monitoring approach, which is capable of sufficiently observing the target behavior in real time, are outlined. Based on the trace data generated by the runtime monitoring approach, the target behavior is visualized/debugged using UML interaction diagrams on the host side. The requirements are discussed in the context of two variants of the runtime monitoring mechanism proposed in this paper, namely, (a) generic software-based and (b) on-chip monitoring.

5.4.1 Classification of Runtime Monitoring

Monitoring can be coarsely defined as the use of probes for producing data traces to help the developer/tester gain insight into the origins of misbehavior in the system under test (SUT). Based on the type of a probe used, runtime monitoring is classified into software, hardware, hybrid, and on-chip monitoring. In software monitoring, additional code is added to the target software to obtain the trace data. This is also termed as software instrumentation. In hardware monitoring a dedicated monitoring hardware is attached to the target system for obtaining the trace data. The use of a combination of additional software and hardware to monitor a target is termed as hybrid monitoring. On-chip monitoring refers to the use of additional built-in, on-chip debugging hardware incorporated in the target to obtain the trace data.

5.4.1.1 Pros and Cons

A drawback of software monitoring is the overhead incurred by executing the additional code, at the target, to obtain the trace data. This incurs memory and time overhead in the target. Interference with the target system's normal operation may arise if the execution of the target software is delayed because of the time spent in the monitoring code. For example, model-based tools, such as [15, 16], make use of software monitoring by instrumentation of the source code to debug, visualize target behavior (using UML diagrams) or execute test cases. Since the debug code is downloaded on the target, the instrumentation overhead is significant. To gain a deeper understanding, consider a simple experiment comprising of application scenarios with 2, 4, 6, and 8 classes in the design model. To visualize the behavior of the target, the instrumented code generated for these examples in [14] shows an increase of 150–250%[1] of source code in comparison with the respective application code size. Clearly, such approaches are not suitable for resource constrained embedded systems. Hence, when using a software monitoring mechanism, it is imperative to minimize the monitoring overhead (e.g., [8]).

The significant advantage of hardware monitoring arises from the nonintrusion benefits obtained by using additional hardware [31]. However, scalability of hardware monitors is affected with respect to monitoring more complex systems [32]. On the other hand, hybrid monitoring makes use of advantages of each approach (i.e., software and hardware monitoring) while at the same time attempts to mitigate their disadvantages.

The key advantage for on-chip monitoring is the presence of an on-chip trace unit, which provides watch points, data tracing, and system profiling for the processor [4]. This can be treated as the major enabling technology, in the future, for target monitoring and testing. A prerequisite is that the underlying processor in the embedded system should support this feature. Whereas, the trace data obtained from the on-chip trace units is in a standardized format (e.g., Manchester encoding) a disadvantage, at this juncture, is the lack of open source/standard tools for communicating the real-time trace data. For example, the real-time trace data from the microcontroller (e.g., MCB1700 evaluation board with Cortex-M3) can be sent to the host only by using proprietary tools (e.g., ULINKpro [6] for Cortex-M3 [4] architecture).

[1]Obtained by measurement.

5.4.2 Time-and Memory-Aware Runtime Monitoring Approaches

In the case of embedded systems, the main factor influencing the usage of a monitoring mechanism is the monitoring overhead [32]. Toward this direction, this section focuses on proposing a time-and memory-aware monitoring approach for debugging and visualizing the target behavior in real time. In this context, the requirements of the two variants of runtime monitoring, relating to the main scope of this paper, are discussed. The two variants are the (a) generic software-based and (b) on-chip (software) monitoring methodology for debugging/visualizing the target behavior on the host side.

5.4.2.1 Software Monitoring

For a software-based monitoring approach to be applicable for deeply embedded systems, with minimal or ideally no overhead, it is imperative that it satisfies the following constraints:

- Generic monitoring routine i.e., independent of the application (size, complexity)
- Minimally intrusive runtime monitoring routine (e.g., a few bytes of memory)
- Modular software monitoring approach, independent of the debug communication interface used
- Minimizing communication overhead between the monitoring routine and the application (e.g., target debugger), which is decoding and interpreting the trace data at the host

With a minimal, generic monitoring routine and bounded, measurable/predetermined overhead (memory, time), the software-based target monitor can be delivered along with the final production code. Now, the additional monitoring overhead (memory, time), which is known beforehand, can be accommodated during the system design phase. This can be achieved by allocating additional resources (memory) or adjusting the scheduling properties (time).

5.4.2.2 On-Chip (Software) Monitoring

Another alternative of runtime monitoring is the on-chip monitoring methodology. However, open source standards for communicating the real-time trace data from the microcontroller to the host, i.e., accessing real-time trace data without the use of proprietary debug adaptors (e.g., ULINKpro [6] for Cortex-M3 [4]) is currently unavailable. On the other hand, on-chip monitoring can be treated as a major enabling technology for the future in the context of minimally intrusive debugging/testing of embedded systems. Hence, the on-chip monitoring approach is chosen as an alternative for visualizing the target behavior in real time, using the model-based debugging approach proposed in this paper.

However, at this juncture, in order to insert the test stimuli/input data to the embedded system and receive the test results using the real-time trace functionality, without proprietary tools, additional hardware, and/or software components are necessitated. In this paper, such an approach (of adding additional software components) is followed in the experimental evaluation for debugging using on-chip monitoring (Sect. 5.5). Hence, this approach is denominated as *on-chip (software)* monitoring in the following section. On the other hand, to apply a memory and time-aware runtime monitoring methodology, such an additional software component, if included, should be a generic monitoring routine with minimal, bounded, and measurable overhead parameters.

An example of on-chip monitoring is available in the recently introduced Cortex-M3 processor/architecture (e.g., used in an evaluation board [27]) that supports real-time tracing using a built-in debug unit called Data Watchpoint and Trace (DWT) and Debug Access Port (DAP). However, to inject the test stimuli from the host computer (e.g., using the test framework approach) additional hardware and/or software components need to be developed.

5.5 Experimental Evaluation

An experimental evaluation based on the two monitoring approaches described above is presented in this section.

5.5.1 Software Monitoring

A prototype implementation of the proposed software-based runtime monitoring approach is described in this section. The aforementioned generic, software-based target monitor can be implemented in programming languages such as C and bundled with the RTOS framework in the embedded system. The target monitor routine is then either invoked by the host or the RTOS (and the generated code from the model) to send and receive debug data respectively. For example, the target monitor routine is invoked by the host (e.g., by the target debugger in the proposed model-based debugging approach) to inject the debug stimuli, in the form of events to the target, i.e., $\texttt{host} \xrightarrow{input} \texttt{target monitor}$.

Similarly, the target monitor is invoked whenever an event is consumed at the target, to send an event-consumed notification to the host, i.e., $\texttt{target monitor} \xrightarrow{result} \texttt{host}$. This trace data is then reconstructed at the host by the target debugger, as UML sequence/timing diagrams to visualize the target behavior in real time.

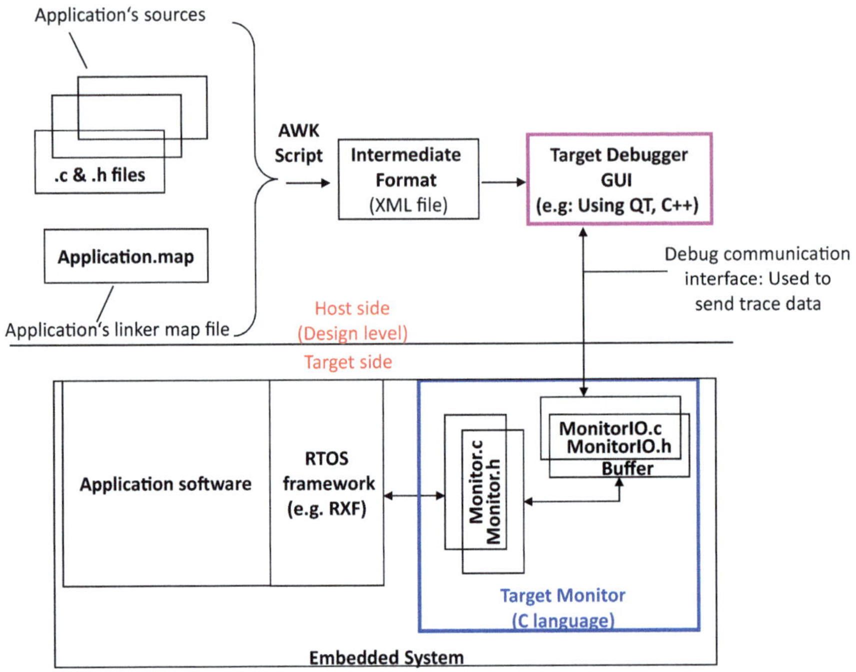

Fig. 5.5 Software-based runtime monitoring at the target side and XML file creation at the host side

Thus, the generic target monitor routine in the prototype (Fig. 5.5.) comprises of two main functionalities, namely, (a) communicating with the RTOS framework and (b) communicating with the host.

The target monitor implementation is modularized based on these two functionalities in *Monitor.h/.c* and *MonitorIO.h/.c* routines respectively. The RTOS framework used in the prototype (OORTX-RXF [33]) comprises of a scheduler that handles the events. The target monitor functionality is used at this point by invoking (a *send* function in) *Monitor.c* in the RTOS framework for consumed events. The module *Monitor.c* in turn uses the functions in *MonitorIO.c* to send and receive data between the host computer and the embedded system via a debug communication interface (Fig. 5.5). *MonitorIO.h/.c* is configurable and implemented based on the APIs and functionalities available in a given debug communication interface (e.g., EIA-232 [1] or JTAG-based). The monitor implementation uses a configurable buffer to handle the trace data.

For example, when an event is processed and dispatched to its respective receiver in the embedded system, *Monitor_sendEvent(unsigned int* pEventData)* function in *Monitor.c* is used to notify the host about the event consumption at the embedded system. This in turn invokes the respective function in *MonitorIO.c* to send the trace data to the host computer, which is decoded by the target debugger in the host computer. The target monitor prototype using a RS-232 debug interface requires a total memory size of approximately 1 Kbyte (1061 bytes ROM plus 135 bytes RAM). A comparison of the target monitor implementation for various debug interfaces and their experimental evaluation is described in Sect. 5.5.

5.5.1.1 Target Debugger

A prototype of the target debugger comprises of a decoding and animation program. The target debugger is implemented in the programming language C++ using the user interface framework Qt [30]. The target debugger decodes and interprets the trace data which is sent via a debug communication interface as seen in Fig. 5.5.

5.5.1.2 XML File Creation

During the compilation of the (application) system code, an AWK script parses the linker map file, source files, header files (of the application), and creates a symbol table data. This is stored in an intermediary format such as an XML file as seen in Fig. 5.5. The decoding program at the host makes use of this XML file (Fig. 5.5) to decode and interpret the incoming trace data which is sent via a debug communication interface as seen in Fig. 5.5. Significant overhead involved in sending trace data back and forth between the host and the target system is avoided by the use of the XML file at the host and predefined frame format for notifications [18].

5.5.1.3 Predefined Frame Format for Notifications

The predefined frame format for notifications, i.e., debug-input data (e.g., inject event) from the host computer and the debug results (e.g., event consumed notification) to the host computer are shown in Fig. 5.6.

The predefined frame format is based on the following design considerations, namely, (a) compactness, (b) minimum number of operations on the target and (c) extensibility. The frame format in the prototype implementation is shown in Fig. 5.6. The "length" field is mandatory, one byte in length and indicates the length of the parameters. The mandatory "command_id" field is also one byte in length. It denotes the command corresponding to the frame sent. The "parameters" field is optional and can be between 0 and 255 bytes in length. It denotes the data about the current command. The minimum length of the monitor frame is two bytes (1 byte each for length and command_id). The frame format for injecting events (from host to target) and the trace data format for sending the event-consumed notification (from target to host) is shown in Fig. 5.6. To inject an event (i.e., debug-input data) the required parameters are the destination of the event, the event to be injected, source of the event, and event parameters (if any). In this frame format, the event, source, and destination values occupy 4 bytes each.

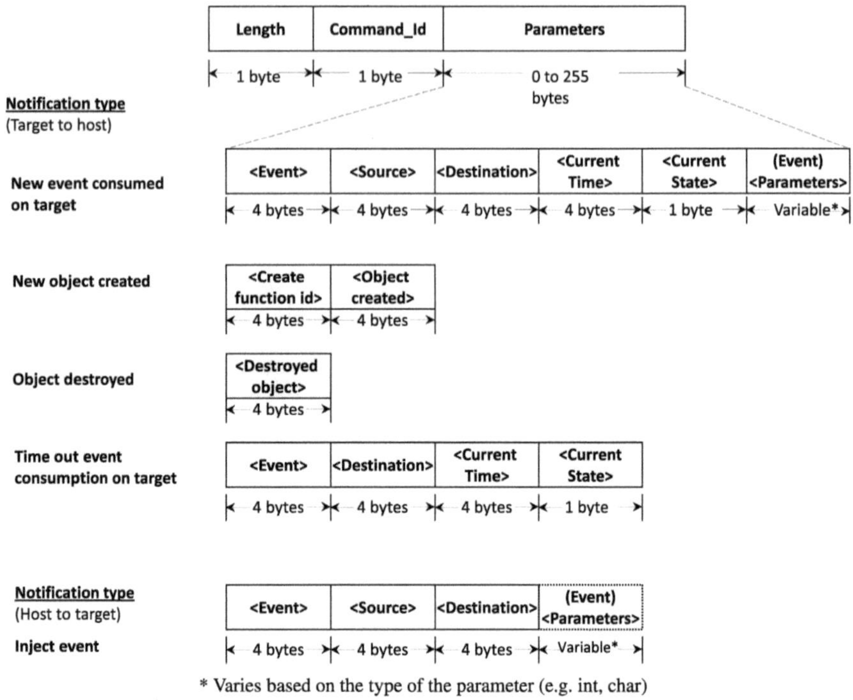

Fig. 5.6 Trace data frame format

For example, to inject an event, *evToggle(int LedNr)* from class *Controller* to class *LED*, to turn on an LED indicated by *LedNr*, i.e., $\texttt{Controller} \xrightarrow{evToggle(LedNr)}$ *LED*, the parameters for the debug-input data in predefined format is *<LED evToggle Controller LedNr>*. However, in the proposed approach, since the object addresses of the above parameters are available at the host computer in the XML file, the debug-input data is *<0X18097098 0X00004374 0X18074897 0X00000001>*. This is received by the *MonitorIO.c/h* routine in the target. The *Monitor.c/h* routine, in turn, decodes, and inserts the corresponding debug-input data to the target. Similarly, the parameters for the debug result for this example in the predefined frame format (Fig. 5.6) are *<evToggle Controller LED CurrentTime ON LedNr>*. In other words, the 21 bytes for the parameters, for the event-consumed notification indicating the debug result, are *<0X00004374 0X18074897 0X18097098 0X00009870 02 0X00000001>*. Note that all events described at the design level, i.e., available in the system code, can be monitored.

Hence by this generic, software-based runtime monitoring methodology, only the debug-input data is injected to the target and the corresponding debug results are obtained as trace data from the target. This implies that the only instrumentation overhead required for debugging in the target is the software-based runtime monitoring overhead. However, note that by optimizing the software monitoring routine and/or further minimizing the memory requirement, the generic software-based run-time monitor can also be part of the final production code.

5.5.2 *On-Chip (Software) Monitoring*

An example of on-chip monitoring is available in the recently introduced Cortex-M3 processor/architecture (e.g., used in an evaluation board [27]) that supports real-time tracing using built-in debug units such as DWT and DAP. The real-time trace functionality in this microcontroller can be used readily with proprietary tools such as μVision [6]. Whereas such proprietary tools cannot be used for inserting the debug stimuli or monitoring the target behavior.

To realize this goal and in order to provide a generic approach toward on-chip monitoring in this paper, a minimal (generic) software monitoring routine is introduced in the target. This can be termed as on-chip monitoring with additional software instrumentation. On the other hand, a trace adaptor (hardware circuit) is necessitated to forward the trace data from the on-chip unit to the host (without the use of any proprietary tools). The aforementioned trace adaptor unit and the generic software instrumentation used in the prototype evaluation are discussed below.

5.5.2.1 Trace Adaptor

Figure 5.7 provides an overall view of the on-chip monitoring arrangement in the prototype. It comprises of a DWT unit that provides support for monitoring data as it is being changed at the target. An additional trace adaptor circuit, with a FIFO buffer, is developed in the prototype. The trace adaptor, as the name implies, is necessary to adapt the trace data from the microcontroller (i.e., serial data stream to UDP data stream) and forward the trace data to the host. This arrangement provides sufficient time to process the data stream (byte-wise) at the FIFO buffer and eventually decode the trace data by the end application at the host computer (i.e., the target debugger in this paper). The DAP debug unit is used to provide support for inserting the input data (debug stimuli) to the embedded system.

There exist two paths for data transfer (Fig. 5.7) between the target and the host (indicated by two different line formattings). Each path is responsible for one functionality, namely injecting the debug input/stimuli to the target and sending the trace data to the host respectively. For example, the debug-input data in the form of events is injected to the embedded system with the aid of the DAP unit, using the JTAG interface. The debug-input data (e.g., regarding an event) comprises of an event, its source, destination and event parameters. The debug data result (i.e., the trace data) from the DWT unit is sent via a Serial Wire Output (SWO) interface, which is part of the on-chip debug unit. This is processed by the trace adaptor circuit for further usage at the host. The trace data indicating the debug results (i.e., an event-consumed

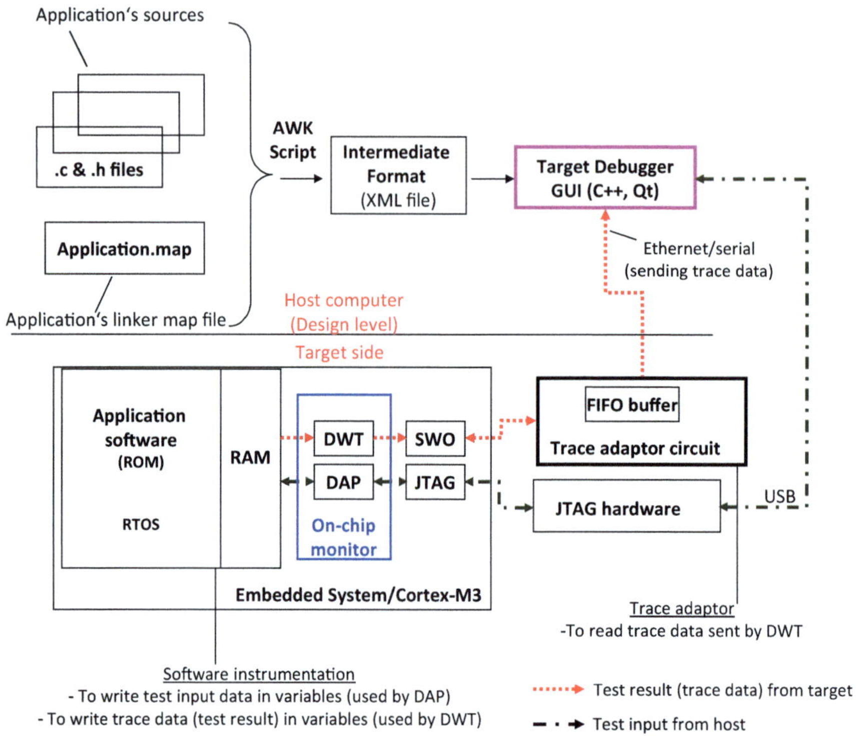

DWT: Data Watchpoint and Trace, DAP: Debug Access Port,
SWO: Serial interface [4],
JTAG: Industry standard JTAG interface [4]

Fig. 5.7 On-chip (software) monitoring arrangement

notification) comprises of an event (consumed on target), its source, destination and event parameters.

5.5.2.2 Software Instrumentation

Software monitoring is necessitated at the target for writing the trace data in data structures, such as predefined (debug) variables of the given application. These variables are monitored by the comparators of the DWT unit. The trace data is sent to the host, when the (debug) variables, monitored by the comparator units, are changed in the target. Thus, the major advantage in the on-chip monitoring approach is that no additional functionality is required to transfer the trace data from the embedded system to the host. Hence, the software instrumentation (in the case of the on-chip monitoring approach) is limited to a few write operation cycles at the target. Then, the trace data, which is now stored in variables/comparators at the DWT unit, is available at the serial interface (SWO) (Fig. 5.7). Therefore, no additional software routines are required for sending the trace data to the host.

Similarly, to inject the debug-input data (comprising an event, its source, destination, and event parameters) sent from the host, software instrumentation is required. The debug-input data is written to the data structures such as predefined (debug) variables monitored by the DWT comparator units. An inject event flag is set after sending the debug-input data to the target. The RTOS used in the prototype detects that an event has to be injected by polling (during idle cycles) the status of the flag (*injectEvent*) at the target, as seen below.

```
if(injectEvent==1){/*check flag, generate event*/
EVT_gen(source, destination, eventId);
injectEvent=0;/*reset flag*/ }
```

Note that *EVT_gen(source, destination, eventId)* is a macro to generate an event in the underlying RTOS framework. Thus, additional software routines to convey the debug-input data from the host or debug results from the target are eliminated by the on-chip debug units such as DWT and DAP in the on-chip (software) runtime monitoring methodology.

5.6 Performance Metrics

Performance metrics such as memory and time overhead of the runtime monitoring mechanisms are discussed here.

5.6.1 Software Monitoring

The monitoring overhead for the software monitoring approach, such as target monitor size, event (bursts) handling, target monitor buffer overflow, time spent in the target monitor routine, and a comparison with the instrumentation overhead in the existing approaches, is presented here.

5.6.1.1 Debug Communication Interface and Target Monitor Size

The software monitoring mechanism introduced in this paper, for debugging, is intended to be independent of the application, its size and complexity. Moreover a modularized implementation of the monitoring routine is proposed in this paper such that the communication of the monitoring routine with the RTOS ((i.e., in *Monitor* routine) and the debug interface (*MonitorIO* routine) are available in two separate routines.

Table 5.1 Memory requirement on target for various debug interfaces using software-based runtime monitoring

Interface	Memory requirement	
	RAM (byte)	ROM (byte)
Generic-EIA-232	112	1290
JTAG-Keil (μVision)	84	1194
JTAG-Lauterbach (Trace32)	84	1396

The implementation of the *MonitorIO* routine is dependent on the debug interface under consideration (e.g., APIs available). In the prototype, the *MonitorIO* routine is implemented for the generic EIA-232 [1] serial interface and two industry standard JTAG-based interfaces, such as Keil-μVision [6] and Lauterbach-Trace32 [25]. The memory requirement for the monitor routine in the prototype for these three debug interfaces is shown in Table 5.1.

From Table 5.1, it is clear that the total memory (RAM, ROM) requirement is approximately 1 KiByte for all the three debug interfaces. Thus, this software-based monitoring routine, which is independent of the application, can also be accommodated in the final production code.

5.6.1.2 Time Spent in the Monitoring Routine

The time spent in the monitoring routine for sending an event-consumed notification obtained by measurement (using a logic analyzer), is shown in Table 5.2. It is clear that the time spent in the monitor routine can be predetermined and is independent of the application and its complexity. The differences in the time spent in the monitor routine for various debug interfaces is based on the implementation and the functionality supported by the APIs for the various debug interfaces [19]. Thus, by the proposed software monitoring technique, the memory (approx. 1 KiByte) and time overhead (in the order of μs), known beforehand, can be accommodated in the earlier phases of the development cycle.

To summarize, target behavior can be visualized online (at the host side), in resource constrained embedded systems without downloading any debug/test harness on the embedded system, using the proposed software-based runtime monitoring mechanism. This is a significant advantage over the existing approaches. The only

Table 5.2 Time spent in software-based target monitor for sending an event consumed notification

Interface	Time in monitor (μs)
EIA-232	74.52
μVision	265
Trace32	16.5

Table 5.3 Number of events handled per interface

Debug interface	Events per second
EIA-232	556
μVision	3770
Trace 32	3268

space requirement in the target is the memory requirement of the software-based runtime monitoring routine.

5.6.1.3 Event (Bursts) Handling

Events consumed at a higher frequency in a short period of time by the target can be termed as event bursts. Since the target monitor implementation depends on the debug interface used, the number of events (and event bursts) the target monitor can handle, also depends significantly on the debug interface under consideration. However, in order to handle the event bursts, the target monitor is implemented with a send/receive buffer interface (Fig. 5.5). The applicability of the target monitor buffer and its dimensioning is again dependent on the debug interface used.

The values shown in Table 5.3 provide a comparison of the number of events that each debug interface can handle theoretically before the use of target monitor send/receive buffers (in our prototype implementation). For example, for the EIA-232 interface, the theoretical maximum number of events that it can handle per second is 556 (115200 [baud rate]/9 [8bit+stop bit]/23 [number of bytes per event-consumed notification]). However, when there is a burst mode in the target system (i.e., number of events per second higher than the theoretical estimation in Table 5.3), this can be handled with the use of a target monitor buffer. Handling of burst-mode data can also be taken over by the APIs provided by the debug interface used. Thus, when there is a consistent burst of events, appropriate dimensioning of the target monitor buffer size and/or selection of a debug interface by the end user is necessary.

5.6.1.4 Target Monitor Buffer Overflow and Real-Time Characteristics of the Target

In the prototype, the target monitor is handled as a lower priority task in comparison with the system tasks. This implies that the target monitor is invoked during the "idle" state of the main loop of the RTOS framework. Let us consider a burst-mode scenario, in which there is a possibility that the target monitor buffer overflows. The target monitor buffer is implemented as a ring buffer. This implies that whenever there is a buffer overflow, the data in the ring buffer could be overwritten. This can lead to a loss of data (notifications about target behavior) stored in the monitor buffer. This is also because of the fact that the target monitor is assigned as a lower priority task and can access the system resources once they are freed by the higher priority (system) tasks. In this case, whenever the target monitor buffer is full, the target debugger is notified about the possible loss of data.

For this scenario there are two possible configuration options, whereby the end user has to compromise between target monitor buffer size and the influence on real-time characteristics of the embedded target. For instance, since the buffer size is configurable, it is up to the end user to allocate a smaller/larger buffer size. As the target monitor implementation is dependent on the debug interface used, the buffer size and its usage is also dependent on the debug interface under consideration. On the other hand, the user could also assign the target monitor as a higher priority task. However, when the end user gives a higher priority to the target monitor (and/or increases the buffer size), he has to compromise between the influence on the real-time characteristics and the loss of target behavior data/notifications.

5.6.1.5 Traditional Versus Proposed Approach–Memory Overhead

The proposed approach has been evaluated for four example scenarios. Similarly, the existing model-based debugging feature (live animation) in the MDD tool [14] was applied to the same evaluation scenarios. Note that the four application scenarios consist of increasing system code (size) and complexity.

For instance, application scenarios 1, 2, and 3 consist of 2, 4, and 8 classes respectively (based on the small "blinky" example [20]). Scenario 4 is based on a more sophisticated case study involving a MIDI system (15 classes). Detailed description of the MIDI system evaluation for the proposed approach is available in [18, 20]. The complexity of the system also varies based on the number of events handled and dependencies on other modules.

The memory overhead incurred (in the target) using both the approaches for the four scenarios (for model-based debugging) are shown in Fig. 5.8. From Fig. 5.8, it is evident that the memory overhead increases with an increase in the application size using the model-based (live animation feature) approach in an MDD tool such as Rhapsody [14]. On the other hand, the size (and the percentage increase) of the target monitor memory footprint is negligible in comparison with the increasing application size as seen in Fig. 5.8 for our proposed approach.

5.6.2 On-Chip (Software) Monitoring

In this case, the only monitoring overhead (time & memory) is that of the additional software instrumentation used to write the test input stimuli/trace data in the debug variables (of the application) monitored by the comparators in the DWT. The additional memory required for the software instrumentation in this approach is approximately 100 bytes (Table 5.4). The time taken to write the trace data for an event-consumed notification (with 23 bytes of trace data denoting the test result), is 360 ns (obtained by measurement using a logic analyzer). The on-chip

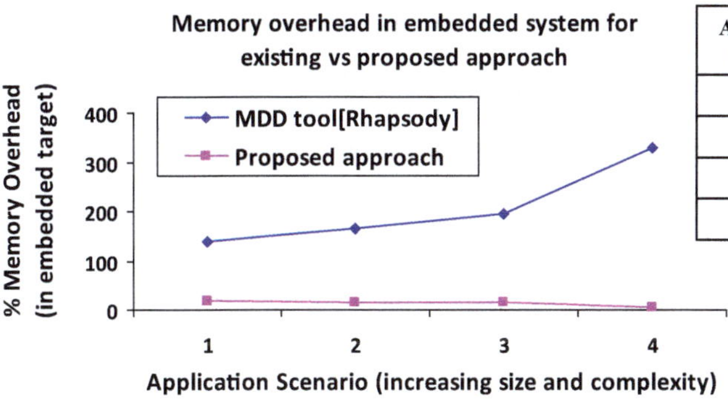

Legend

Application Scenario	Number of classes
1	4
2	6
3	8
4	20

Fig. 5.8 Memory overhead (in target) for model-based debugging of various application scenarios

Table 5.4 Memory requirement on target for software instrumentation

RAM (byte)	ROM (byte)
24	74

monitoring mechanism is also independent of the application and its complexity. Thus, the overhead parameters for executing the test cases using this monitoring methodology (memory approx. 100 bytes and time = 360 ns) is also known and measurable beforehand.

5.7 Discussion and Evaluation

Debugging RTESS is a challenging task in comparison with debugging desktop systems. While there are some traditional and model-based tools for debugging RTESS, these have limitations. Model-based debugging techniques in the existing tools usually involve dynamic source code instrumentation. This instrumented code increases with the increase in the application size and necessitates sophisticated debug interfaces. The protocol and performance overhead incurred during debugging could also result in modifying the temporal behavior of the embedded system. All these factors make the existing model-based debugging techniques unsuitable for the memory-size constrained RTESS.

In order to overcome the aforementioned limitations, a model-based debugging methodology for small RTESS was outlined in this paper. Using the proposed methodology, RTESS behavior can be visualized in real time using UML sequence and timing diagrams. Some salient features in the proposed approach, which overcome the limitations of the existing approaches, are discussed below.

5.7.1 Salient Features in the Proposed Approach

- Dynamic source code instrumentation is eliminated with the introduction of an optimized monitoring software routine in the target monitor (implemented in the programming language 'C'). The target monitor (library) is bundled with the RTOS used (RXF [33]). The target monitor is now independent of the application, its size, complexity, and source code. The target monitor occupies approximately 1 KiByte of memory, which is accommodative for small embedded platforms. Moreover, because of its size, the target monitor can be bundled along the final production code as well.
- In addition to the optimized target monitor size, the information exchange between the target debugger and the target monitor is handled via a custom-defined protocol. The protocol design is extensible, compact, and requires a minimum number of operations. For example, the minimum frame-size for the protocol is 2 bytes and an event consumed notification requires 23 bytes of data. The frame format of this protocol is described in detail in Sect. 5.5.
- A major factor influencing the real-time behavior of the embedded system is the communication overhead between the target system and the host computer (i.e., huge debug data being sent back and forth between the target and the host computer). In order to minimize this, an AWK script parses the source files, header files, and linker map file (for a given project) and creates a symbol table at the host computer in our approach (Figs. 5.5 and 5.6). This symbol table enables identifying each element in the system code, such as class, instance, event, etc., by an ID. This symbol table data is stored in an intermediary format, such as an XML file, at the host computer. On receiving the trace data from the target, the target debugger decodes the trace data with the aid of this XML file. The trace data from the target is translated and the animation program in the target debugger GUI re-constructs the target behavior in the form of UML sequence and timing diagrams in real time.
- Based on the runtime monitoring mechanisms discussed above, it can be stated that the proposed techniques are time-and-memory aware (supported by the performance metrics discussed in Sect. 5.6). This is primarily because, the overhead (time & memory) introduced by the two variants of the monitoring technique is measurable beforehand, minimal, bounded, and independent of the application. On the other hand, tools such as [14] introduce unbounded and variable overhead for debugging and/or visualizing the target behavior (using UML diagrams) in real time for different application scenarios. These features eliminate the risks due to the change in program behavior before and/or after debugging the system code. Therefore, the proposed time-& memory-aware runtime monitoring mechanisms provide a significant improvement over the existing techniques.
- The proposed mechanisms also address the aspect of scalability and applicability for resource constrained targets and industrially relevant examples. For example, the solution proposed in this paper already concentrates on resource constrained embedded systems, thereby addressing the question of scalability. For industrial applications involving several complex interactions and entities, the overhead

parameters can be accommodated in the earlier phases of the development cycle. A robust implementation of the software instrumentation may further reduce the overhead parameters.

5.8 Conclusion

While monitoring/testing an embedded system and acquiring the trace data, one often faces the so-called "Heisenberg's effect": *Inspecting a system tends to influence the system's behavior* [10]. Whereas a time & memory-aware runtime monitoring mechanism is introduced in this paper, a runtime monitoring mechanism which introduces (ideally) no overhead in the target is an ambitious goal. Therefore, while employing a generic software-based runtime monitoring approach, the goal should be to minimize the monitoring overhead as far as possible. An example of this approach is discussed with a prototype in this paper. For embedded systems with microcontrollers supporting the real-time trace functionality [10] (i.e., on-chip debug units), the overhead parameters can be further minimized using an on-chip mechanism such as the one introduced in this paper. When the nature of the embedded software described requires the system to meet real time requirements in debugging/testing mode, the overhead parameters from monitoring can be included in the earlier stages of the development cycle. By doing so, the influence on the real-time characteristics of the embedded system because of the overhead introduced by monitoring can be eliminated.

Application of the proposed time-& memory-aware runtime monitoring, to industrial case studies, adding UML state chart diagrams in the target debugger GUI for visualizing the target behavior and evaluation on other target platforms are some items for future work.

Acknowledgements This work was supported by a grant from BMWi-ZIM, DAAD, and industrial partner Willert Software Tools GmbH. We would like to thank the project teammates at Willert Software Tools GmbH and UAS-Osnabrueck for their cooperation.

References

1. Axelson J (2007) Serial port complete: COM Ports, USB Virtual COM Ports, and Ports for Embedded Systems, 2nd edn. Lakeview Research
2. BridgePoint UML Tool (2016) http://www.mentor.com/
3. Bunse C, Gross H-G, Peper C (2007) Applying a model-based approach for embedded system development. In: 33rd EUROMICRO conference on software engineering and advanced applications
4. Cortex-M3 Processor (2016) http://www.arm.com/
5. Delgado N, Gates AQ, Roach S (2004) A taxonomy and catalog of runtime software-fault monitoring tools. IEEE Trans Softw Eng 30(12):859–872
6. Embedded development tools (2016) http://www.keil.com/

7. Enterprise Architect tool (2016) http://www.sparxsystems.com/
8. Fischmeister S, Lam P (2010) Time-aware instrumentation of embedded software. IEEE Trans Ind Inform 6(4):652–663
9. France RB, Ghosh S, Dinh-Trong T, Solberg A (2006) Model-driven development using UML 2.0: promises and pitfalls. Computer 39(2):59–66
10. Ganssle J (2008) The art of designing embedded systems, 2nd edn. Newnes
11. Graf P, Muller-Glaser KD, Reichmann C (2007) Nonintrusive black- and white-box testing of embedded systems software against UML Models. In: Proceedings of the 18th IEEE/IFIP international workshop on rapid system prototyping, pp 130–138, Washington, DC, USA, 2007. IEEE Computer Society
12. Harmon T, Klefstad R (2007) Interactive back-annotation of worst-case execution time analysis for Java microprocessors. In: 13th IEEE international conference on embedded and real-time computing systems and applications, RTCSA 2007, pp 209–216
13. Huang X, Seyster J, Callanan S, Dixit K, Grosu Radu, Smolka Scott A, Stoller Scott D, Zadok Erez (2012) Software monitoring with controllable overhead. Int J Softw Tools Technol Transf 14(3):327–347
14. IBM Rational Rhapsody Developer, Ver 8.2 (2016) http://www.ibm.com
15. IBM Rational Rhapsody Test Conductor Add-on (2016) http://www.btces.de/
16. IBM Rational Test RealTime (2016) http://www-01.ibm.com/software/awdtools/test/realtime/
17. Iyenghar P, Wuebbelmann J, Westerkamp C, Pulvermueller E (2013) Model-based test case generation by reusing models from runtime monitoring of deeply embedded systems. IEEE Embedded Syst Lett 5(3):38–41
18. Iyenghar P (2012) A test framework for executing model-based testing in embedded systems. PhD thesis, University of Osnabrueck
19. Iyenghar P, Pulvermueller E, Westerkamp C, Uelschen M, Wuebbelmann J (2011) Model-based debugging of embedded software systems. Gesellschaft Informatik (GI), softwaretechnik (SWT), pp 31–33
20. Iyenghar P, Westerkamp C, Wuebbelmann J, Pulvermueller E (2010) A model based approach for debugging embedded systems in real-time. In: Proceedings of the tenth ACM international conference on Embedded software, EMSOFT '10, NY, USA
21. Jiao Y, Zhu K, Yu Q, Wu B (2006) Towards model-driven methodology: a novel testing approach for collaborative embedded system design. In: 10th International conference on computer supported cooperative work in design, 2006. CSCWD '06, pp 1–5
22. Karsai G, Sztipanovits J, Ledeczi A, Bapty T (2003) Model-integrated development of embedded software. Proc IEEE 91(1):145–164
23. Kashif H, Mostafa M, Shokry H, Hammad S (2009) Model-based embedded software development flow. In: 4th International design and test workshop (IDT), pp 1–4
24. LabVIEW System Design Software (2016) http://www.ni.com/labview/
25. Lauterbach-Microprocessor development tools (2016) http://www.lauterbach.com/
26. Matlab and Simulink (2016) http://www.mathworks.com/
27. MCB1700 evaluation board (2016) http://www.keil.com/mcb1700/
28. Object Management Group (2016) http://www.omg.org
29. Plattner B (1984) Real-time execution monitoring. IEEE Trans Softw Eng SE-10(6):756–764
30. Qt. User interface framework (2016) http://qt.nokia.com/
31. Tsai JJP, Fang K-Y, Chen H-Y, Bi Y-D (1990) A noninterference monitoring and replay mechanism for real-time software testing and debugging. IEEE Trans Softw Eng 16(8):897–916
32. Watterson C, Heffernan D (2007) Runtime verification and monitoring of embedded systems. IET Softw 1(5):172–179
33. Willert Software Tools GmbH (2010) http://www.willert.de/

Chapter 6
A Mechanism for Monitoring Driver-Device Communication

Rafael Melo Macieira and Edna Barros

6.1 Introduction

The use of electronic embedded systems for general or multipurpose applications has increased substantially. IoT and cyber-physical systems are some examples in which embedded systems require more flexibility for processing a different kind of applications and communication and control protocols.

The need for this flexibility with high processing power makes the modern embedded systems, composed even more by multiple general purpose processors and several software layers, extremely complex. Additionally, these systems also require critical and extremely tight constraints as task deadlines and area and power consumption limitations, for example.

The whole design can become very complicated in the case of cyber-physical systems (CPS). Cyber-physical systems integrate computation with physical processes. They include embedded computers, networks monitor, and physical processes controllers usually with feedback loops where physical processes interfere and guide computations and vice versa [12]. Thus, it can be tough and time-consuming to construct a reliable system without appropriate techniques and tools.

For CPSs, either the hardware and the software design are extremely complex. There is a trend requiring them to be highly dependable, reconfigurable and, in some cases, certifiable [3]. Thus, it is almost impossible to construct a reliable system without appropriate techniques and tools. If even the design of platforms with single conventional general purpose processors needs time and means to produce a reliable system, for emergent technologies like NoC-based MPSoCs and CPSs this need is even greater.

Besides, it is common to see CPSs applied to critical environments, such as mass transportation, power plants, or medical equipment, which involve several lives and

R.M. Macieira (✉) · E. Barros
CIn - Informatics Center, UFPE - Federal University of Pernambuco, Recife, Brazil
e-mail: rmm2@cin.ufpe.br

E. Barros
e-mail: ensb@cin.ufpe.br

D. Lettnin and M. Winterholer (eds.), *Embedded Software Verification and Debugging*, Embedded Systems, DOI 10.1007/978-1-4614-2266-2_6

expensive infrastructure. So even being a challenge to design such systems, the existence of flaws in its execution is unacceptable.

In addition to the failure issue, another motivation in optimizing embedded software designs is the reduction of the time-to-market. The market window requires the manufacture of electronic devices in short time intervals, which requires the reduction of time for modeling, simulation, and verification of the hardware/software system. Thus, techniques and computational tools must be developed to allow costs and design time reduction.

Trying to cope with this productivity gap, also increasing the reliability of complex embedded systems, the author of this document requesting the scholarship proposes a nonintrusive technique for detecting critical errors and behaviors patterns in hardware-dependent software during their development or their update on runtime environment.

Detecting critical behaviors during the design phase prevent releasing faulty devices. The costs of repairing a software flaw during maintenance are 500 times higher than a fix in an early design phase [4]. However, sometimes the devices suffer update, and their whole system must be ready for detect e react to critical situations, being fault tolerant. Several research works [7, 10, 11, 20, 22] show the importance and viability of fault tolerant systems, isolating failure and taking over the system, leading it to a safe state. But to take advantage of this technique, the system must be ready to identify or predict critical behaviors.

So, the main idea of the proposed approach is monitoring on runtime every access made by the embedded software to system platforms resources and, based on their modeled reference standard behavior, implemented as a finite-state machine, checking if embedded software is respecting critical properties of the platform.

The front-end of this approach is a set of high-level specifications written in a domain-specific language (DSL) called TDevC, that contain structural and behavioral descriptions of the devices and the whole platform, and the communication between the devices and the embedded software. This language has been developed since 2009 and has been extended to allow more complex specification of registers accesses constraints as well as allowed sequences of registers accesses, as can be seen in [15, 16], until this current syntax and expression power.

Added to this description, a set of assertions, described through linear temporal logic (LTL) formulae, are also specified.

The architecture of the proposed mechanism includes a monitoring module called MDDC containing a finite-state machine set (FSM-set), which can fire the assertions, based on the sequence of accesses to the device. Synthesized to SystemC models and SystemVerilog components, these modules must be integrated into a platform model (either a virtual prototype or FPGA prototype).

Some experiments performed with simple and complex devices and environments has shown the feasibility of this approach. The monitor module controlled all assertions. An interesting characteristic observed during the experiments was the ability of specifying and catching cross-layer behaviors in the software stack.

In the further sections, this paper discusses related works, describes how the language performs the checking, details the language TDevC and shows the experiments performed and theirs results. In the end, it concludes the work, discussing a little bit about the contributions.

6.2 Related Works

Several approaches try to cope with correctness issues related to embedded software. Different ways of looking at the problem provide a different method to reduce the presence of bugs, increasing the reliability of embedded systems.

Correct-by-construction is one method used to decrease the occurrence of errors. The bet of this kind of technique is the reduction of software coding. Through a high-level specification, such as domain-specific languages (DSL), a synthesis tool automatically generates the device driver code. Some approaches, as described in [1, 14, 18], proposes the automatic generation of device drivers. However, as a common drawback of this technique when applied to the synthesis of device drivers, the part of the software dependent of the hardware (the driver) (Hardware dependent Software HdS) is not generated completely. Thus, in these cases, always there is a need for coding some part of the generated software by a human being, which can be a source of failure. Besides, this kind of technique looks only for the lowest level of the software stack, however sometimes the systems' failure are in a higher layer.

Another way to increase the reliability of the HdS layer is betting on system's resilience [7, 10, 11, 22]. Some approaches focus on the fault tolerance of the drivers, isolating failure and taking over the system, leading it to a safe state. Despite being effective, this kind of technique commonly generates a considerable overhead in the system's execution and, in spite of isolating the failure, the fault in some cases continues existing.

Several approaches deal with faulty device drivers performing formal or semi-formal verification during the development phase [2, 5, 6, 13, 17, 21].

However, the works proposed by Lettnin and Behrend [5, 6, 13], for example, need some code instrumentation, which sometimes can decrease performance or change timing behavior. Besides, applied only for high-level models, for cycle-accurate simulation, the checking simulation can be very time-consuming. Even proposing the reduction regarding verification time, these works [5, 6] do not provide any cycle-accurate verification, which limits the validation of significant embedded software constraints.

Reinbacher, in [17], proposes a microcontroller embedded software runtime testing environment based on accesses to memory-mapped areas. However, this approach needs the software binary to extract the variables addresses, what makes this approach extremely dependent on the software stack. The result of the validation as well as the configuration of the environment must be analyzed and controlled by a host machine through a USB port, making difficult its application for resilience and adaptation of systems.

Another point of concern related to reliable systems that none of the approaches presented above took into account involves software maintenance. Any software layer modification may imply in an environment verification modification. Software variables names and addresses may change and libraries, frameworks,and operating systems that surround the HdSs also may change. These modifications may imply in binding properties changes. It is important to highlight that with an independent relationship with the software layer implementation, its source code and its binaries, the proposed technique applied to on-the-market devices can prevent them from faulty and unreliable software updates. Additionally, it and can aggregate basic resilience to the system in case of failure of devices, once this nonintrusive validation environment runs on the same platform of the device and it can provide feedback to its system.

6.3 Proposed Approach

As mentioned in Sect. 6.1, this work proposes a technique for monitoring and checking the correctness of embedded software and the communication between them and the peripherals of the platform.

To perform the monitoring, the proposed approach defines a strategy for synthesizing specific assertion monitors, the MDDCs, which, during the execution of the platform, snoops the communication interconnection between the processing element and the peripherals and verify the veracity of system assertions, in order to guarantee the correct and reliable platform's execution.

The abstract idea of the use of the MDDC can be seen in Fig. 6.1. The sequence of frames in the Figure, starting from the 1 to the 6, simulates the use of a platform containing one CPU, two peripheral named as *Dev #1* and *Dev #2*, and the MDDC module. At the frame 1, the whole platform in a previous stage can be seen, before performing any communication with the devices. At the frame 1, it can be seen the communication interconnection, which in this example is a bus, connecting all the components. Additionally, the interconnection links the MDDC module to the CPU and the bus and directly connected to the CPU through an interrupt line.

From the frame 2 to 6 there is a sequence of accesses to devices. Accesses means readings and writings to registers of peripherals. For this hypothetical example only *Dev #1* is under verification. Thus, the frame 2 shows that when the CPU makes a writing access to the *Dev #1*, the MDDC module recognizes this writing as a legitimate access to the verification and accepts it as an input. Following, at frame 3, the CPU performs another writing, but now to the *Dev #2*. However, *Dev #2* is not an under validation device and, then, the MDDC module ignores that access as an input.

The next access showed in Fig. 6.1-frame 4 is an example of a reading access to the *Disp #1*. The MDDC module catches the access, accepts it and waits for the response with the data requested by the CPU. The MDDC module captures that returned data, as showed at frame 5, and adds it to the previously mentioned reading access.

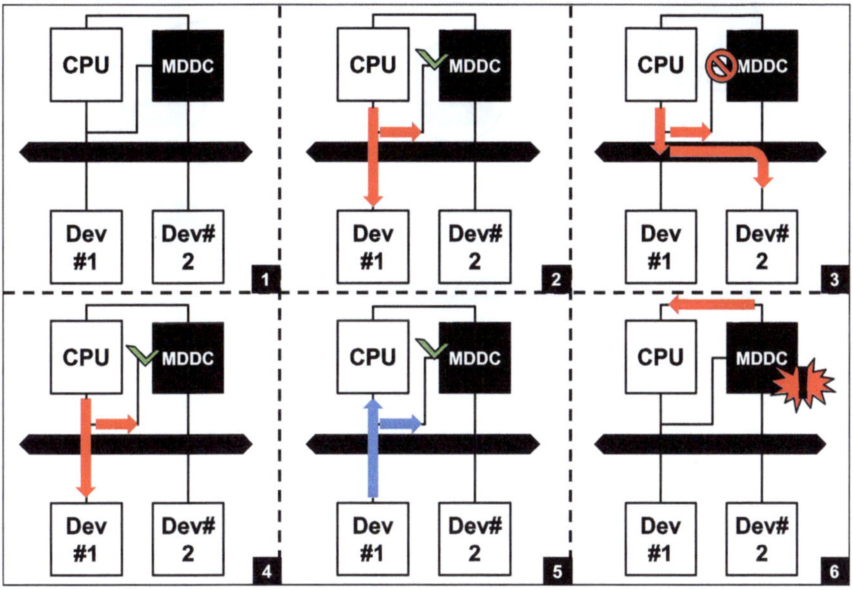

Fig. 6.1 General basic approach

To finish the example, the Fig. 6.1-frame 6 shows a case of a detection of an assertion. Supposing that the reading access performed at frame 5 has triggered an assertion, the MDDC module reports the event and, in this case, interrupts the CPU.

It is important to highlight that the interrupt line is one of the several possible ways to report an assertion detection by the MDDC module. In the example showed in Fig. 6.1, the CPU can use information about the assertion to make any decisions in the software execution. Despite it being an intrusive technique, it is widely used for resilient, self-aware and adaptive systems, such as Cyber-Physical Systems (CPSs).

Once we described how the MDDC module works together with the platform, now it will be explained how it uses the accesses to monitors the behavior specified through assertions. The MDDC module includes two state machines for supporting assertions monitoring: the Hierarchical finite-sstate machine with data assignment(HFSM-D) and the Büchi Automaton (BA). Only one MDDC module is enough for monitoring the communication between one processing element and several devices. Thus, an instance of an MDDC module contains information about each device under validation, in which, the main information is related to the devices' execution basic behaviors, represented by the HFSM-D state machine, reflecting the device response, as a reference model.

The Fig. 6.2 shows a simplified example of an HFSM-D state machine, showing how it reflects the execution behavior of a device under validation, and demonstrates how the writing to this device, as the example of Fig. 6.2-Frame 2, would act internally in the monitor.

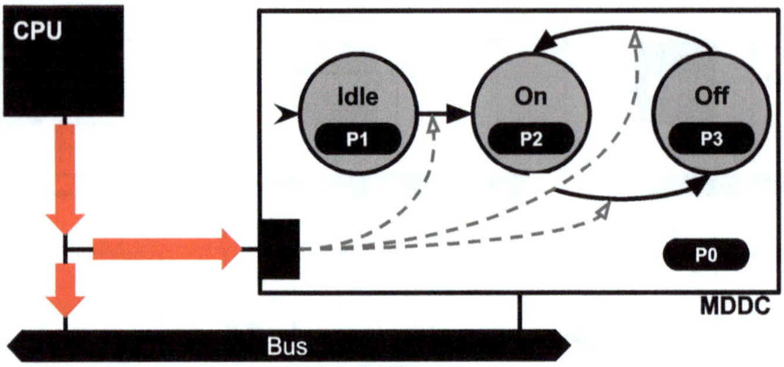

Fig. 6.2 Approach flow in progress

Once the MDDC module accepts it as valid, the access is translated into transactions in an HFSM-D state machine. Based on the example of Fig. 6.2, the initial state of the state machine is the state *Idle*. Thus, as the HFSM-D state machine is a standard behaviors model of the execution of the under validation device, it is expected that the device is initialized in an equivalent state *Idle* internally.

Referencing again the Fig. 6.2, one can see that each state contains a block representing the set of properties that must be respected at the state. There is a set of properties (*P0*) out of any state, where the properties belonging to the set *P0* are global. The following section includes a more detailed description of scopes and hierarchies of both properties and states.

These properties, which states must comply, are implemented using Büchi Automata (BA). They are formal representations of linear temporal logic properties [19], i.e., properties that must be met over time. Based on temporary events the BAs perform state transitions always after updates of states and values in the HFSM-D state machine. Temporal events are events that occur in time slots, not necessarily in constant time. Examples of temporal events: clock in integrated circuits, banking transactions, sending and receiving network packages, access to devices, and so on. In this approach, the temporal events are the accesses made by processing elements to devices and clock ticks in the case of real-time properties.

This is the overall simplified goal of the proposed approach. However, to implement and to integrate the MDDC module using a general purpose language or an HDL is not a simple task and can consume a good amount of design time, since each MDDC is specific to the target platform.

Moreover, the implementation of a validation environment from scratch using general purpose languages may make it highly susceptible to errors, since the manual coding of functional and nonfunctional design constraints needs to be taken into account. Thus, it is ideal that the verification environment can be specified using a high-level abstraction and taking into account only the functional requirements of the design.

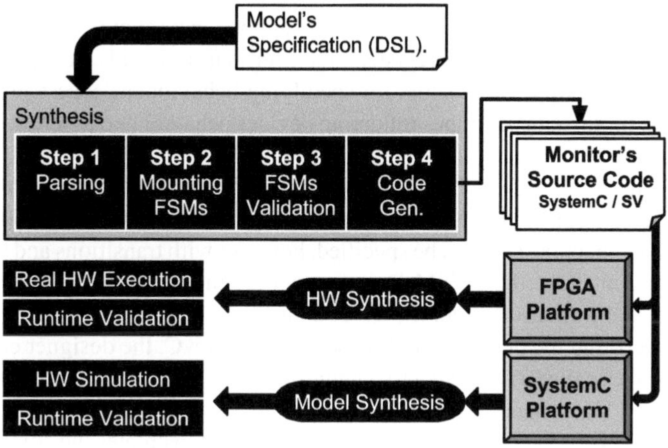

Fig. 6.3 Approach's main flow

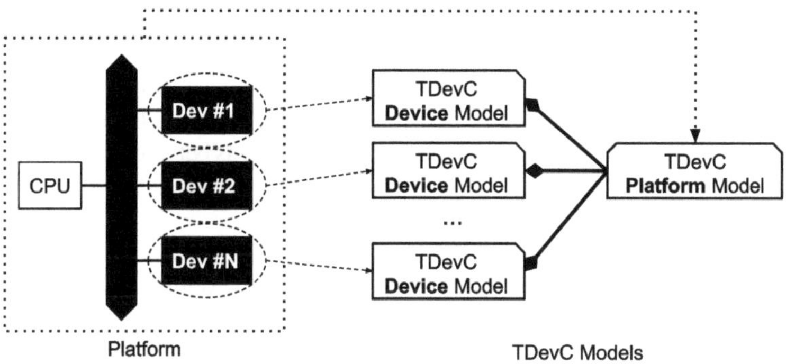

Fig. 6.4 TDevC models' diagram

Thus, this approach proposes a technique for developing a verification and execution environment through the sequence of steps that can be seen in the flow shown in Fig. 6.3.

The proposed flow starts with the specification of high-level abstraction models of structures and behaviors of the under validation device and the whole platform. These models are described in a DSL called TDevC, proposed in this paper.

As it can be seen in Fig. 6.4, there are two types of TDevC models: device model and platform model. Using the device model the system designer specifies the structures and behaviors of each device in the platform, and can specify the assertions related to each single device.

Through the platform model, the system designer can specify custom behaviors of the whole platform execution and system assertions based on shots of devices' assertions.

A TDevC device specification mechanism supports in its syntax constructions the description of structures such as registers, their formats and fields, data patterns, and masks. It also supports constructors for specifying behaviors as access protocols to the registers, finite-state machines reflecting devices behavior and assertions through temporal properties.

On the other hand, the TDevC platform specification approach supports only behavioral constructors. Besides the declaration the devices under validation, an HFSM-D state machine also can be specified, however with transitions and assertions triggers different from the HFSM-D state machine modeled in the TDevC device models. More information about the TDevC language will be given in Sect. 6.5.

After specifying the device models described in TDevC, the designer can use the proposed toolset called *TDevCGen* for synthesizing the monitor module that includes the mentioned state machines. Including four steps, the synthesis begins with the parsing of the TDevC specifications. The TDevCGen then creates the control lists of structural and behavioral elements and then in step 2, translate the models described in TDevC to an intermediate format, assembling in memory the state machines in a hierarchical manner and linking its states to the structural elements associated with their transitions.

With the intermediate format in memory, the toolset can then perform the third step by validating the specified model. This validation aims to check for inconsistencies and nondeterminism in the state machine and contradictions between the assertions of states, their immediate descendants and between the device models and the platform model.

With the validated model, it follows to step 4. The TDevCGen then generates the source code in SystemC or synthesizable SystemVerilog. With its full source code, the monitor can be integrated into the platform. Currently, the integration in done by the designer manually.

The integration step consists in connecting the MDDC generated into the target platform. The physical ports (low-level of abstraction) or SystemC functions (high-level of abstraction) that provide to the masters devices the accesses to memory-mapped addresses, available through platforms libraries, are snooped or encapsulated. So, the data handled by these snooped or encapsulated ports and functions are sent to the snooping port of the monitor.

After manual integration, the MDDC module is ready to intercept all data sent from a master device to the peripheral devices during the execution of embedded software and determine whether the accesses represent the desired behavior or not.

All data exchanged between the master device and the device under validation are assessed and, depending on the type of access, the related address and read or assigned value, a state transition in the hierarchical machine can be triggered. So with the hierarchical state machine reflecting the standard behavior of the device under validation and associating assertions to the states, the monitor can detect any time during system execution the state of peripherals and the occurrence of assertions associated with the current state.

Importantly, in addition to the monitor being a spy mechanism used to snoop the communication between master and slave elements within the platform, it is also seen as a slave device. Therefore, the processing element can configure it and, occurring a specified behavior fired by an assertion, requests the MDDC module for more information about the event. As said before, this is interesting not only for the detection of faulty behaviors and debugging software but also for any decision-making in the face of their occurrence.

6.4 Definition of the HFSM-D State Machine

This approach makes use of state machines to represent the basic behaviors of the device under validation and the system assertions.

The HFSM-D state machine is used to represent the standard behavior of the device under validation, from the embedded software. On the other hand, the assertions are translated to Büchi automata. This section will explain in detail the HFSM-D state machine. The definition of Büchi automata can be found in [8].

Each state of the HFSM-D state machine represents an execution possible state of the device. To support different abstraction levels, a hierarchical description is allowed. A state can contain a complete substate machine, including more states, known as *child states*. The deepest in the hierarchical level, the more granular and detailed is the specification of the device execution.

The root state of an HFSM-D state machine called Global state, is always a unique state, ancestral of all the states in a hierarchical machine description. Within each parent state, that is, those who, in the perspective of hierarchy, are not a leaf state, there are one or more parallel and separate substate machines that reflect concurrent behavior. Within a state is said that each parallel substate machine is inside an orthogonal region. Thus, each non-leaf state can contain one or more orthogonal regions containing, each one, a separate state machine and parallel execution. These regions inherit the same principles introduced by Statecharts [9].

The Fig. 6.5 shows an example of an HFSM-D state machine proposed in this work. In this example it can be seen that there are 17 states, called *s1* to *s16*, plus the Global State *g*. Still in the same example we can see that there are four orthogonal regions, two of them (*o1* and *o2*) belonging to the state *g* and the other two, to the state *s3*.

Orthogonal regions are explicitly specified in the model to prevent the occurrence of direct transitions between different substate machines. As distinct execution lines, joints in the transition flows of different parallel substate machines and different hierarchy levels are not allowed.

Using the Fig. 6.5 to make this explanation clearer, the transition from the state *s12* to the state *s13* may occur at the same time of the transition between the states *s8* and *s9*, for example, depending on the transitions' trigger.

Transitions are fired based on a Boolean expression, for devices models, and based on device assertions, for platform models. The atomic propositions of devices models are boolean expressions, variables value comparison, current execution states,

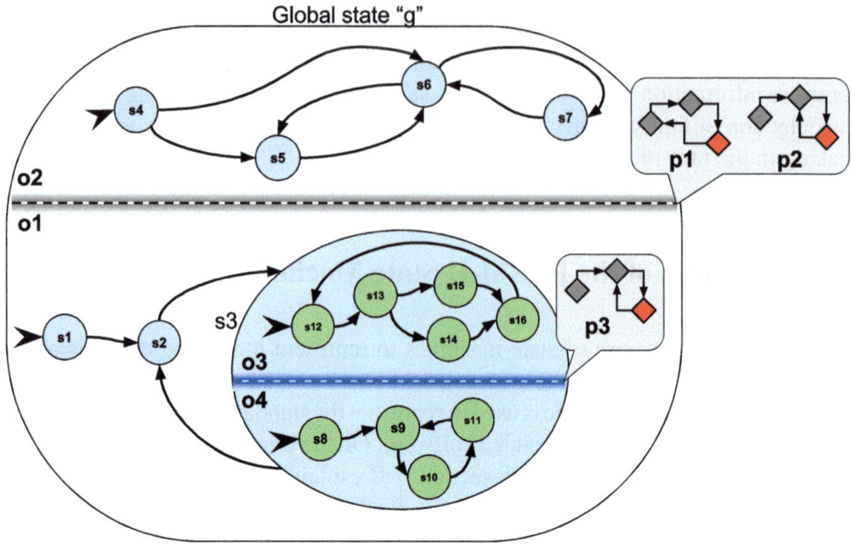

Fig. 6.5 Hypothetical example of a HFSM-D

accesses information, and time delays. All transition expression of a current state is tested if it is truly every time an access occurs or when a time delay ends.

During the synthesis of an HFSM-D state machine, some transitions are automatically created depending on their types. As a simplification, each state transition can be defined as *exit points* or *entry points*. The exit point is treated as normal transitions, starting at the state that has specified it and reaching a target state. On the other hand, the specification of entry point transitions informs to the synthesis tool that every state of that orthogonal region must have a transition to the state which has declared the entry point transition. This situation is common during devices resets, for example. Normally all transitions to a reset state in the orthogonal region have the same trigger. So, it is not necessary to specify one transition for each state with the same trigger and the same target state. The target state itself specifies it.

Even well structured and with some transitions automatically inferred, a lot of inconsistencies can remain in the generated model. Thus, the toolset fulfills a sequence of model checking. The first checking looks for transitions to different state machines, i.e., different orthogonal regions. As said before, each orthogonal region contains only one substate machine. Thereby, their states only can have transitions to states in the same substate machine. Again using the Fig. 6.5, the state *S12* can only have transitions to the states *s12* to *s16*, but never to the states *s8* to *s11*, for example. This checking analysis prevents the merging of sub-state machines execution lines.

Another inconsistency that must be verified is the nondeterminism. Using a theorem prover, every transition of a certain state is compared with the others transitions of the same state. The theorem prover resolves the Boolean logic expression and checks if at least two transitions can occur at the same moment. The occurrence of a transition means that the Boolean logic expression is true in a transition event for a certain state.

The last checking performed by the toolset is to look for properties contradictions. However, before talking about it, it is essential to clarify some information about these properties. Expressed in LTL formulae and translated to Büchi automata applying the technique proposed by [8], these properties represent behavioral characteristics of the execution that can be identified and signalized by the validation environment. Note that, if every state may specify a set of properties, the scope of them is limited to that state and their child state. These temporal properties only must be held to be signalized when the execution is in that state and, consequently, in its child states. This elements are represented in the figure by symbols *p1*, *p2* and *p3*. The properties *p1* and *p2* are directly associated with the state *g* and hence indirectly associated with all other states of the HFSM-D state machine. The *p3* is directly related to state *s3* and indirectly associated to its child states *s8*, *s9*, *s10*, *s11*, *s12*, *s13*, *s14*, *s15*, and *s16*. For all other states, *p3* has no effect.

It is important to clarify that, for the state *s3* and its child states, both properties *p1* and *p2* as *p3* must be considered. Because of the hierarchy, parent's properties are also valid within the child states.

To make clearer the commented hierarchy levels relationship, Fig. 6.6 shows a hierarchical view of the example showed in Fig. 6.5. In this figure, it can be seen the Global State *g* at the top and consequently at level 0. Below, at level 1, are its two orthogonal regions (*o1* and *o2*) and their child states (*s1* to *s7*). Following, at level 2, the orthogonal regions of *s3* (*o3* and *o4*) and its child states (*s8* to *s16*).

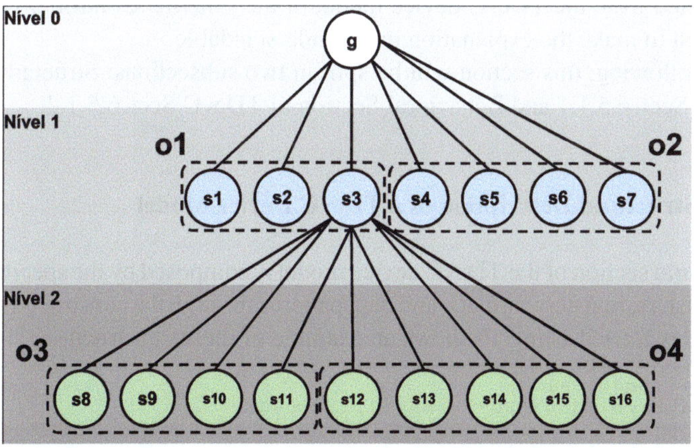

Fig. 6.6 Hierarchical point of view of the HFSM-D of Fig. 6.5

6.5 The TDevC Language

As mentioned in Sect. 6.1, the front-end of the proposed approach is a high-level specification described in a DSL language called TDevC. This language has been developed since 2009 and is being adapted, as it can be seen in [15, 16], to have more expression power to capture system assertions and constraints.

To maintain the central features of a high-level specification mechanism, the TDevC permits to describe the structural specification of the device, devices memory access protocols, devices behaviors, and platform behaviors. The two last as it will be explained in the following paragraph are expressed through syntax constructions used for defining the hierarchical finite-state machine with data assignment (*HFSM-D*).

As explained in the previous section, there are two types of TDevC models. Thereby, despite very similar, they have different syntax. So for a better organization of this paper, this document will split the explanation in two subsections: *TDevC Device Model Section* (6.5.1) and *TDevC Platform Model* (6.5.2).

6.5.1 TDevC Device Model

The TDevC device models describe the behavior of the peripheral devices of a given platform, based on their fundamental structural components. Thus, these models include language constructors to support the specification of both the structure and the behaviors.

Thus, this section will describe these constructions. For each one of them, examples extracted from the TDevC device model of the Ethernet Controller DM9000A will be used to make the explanation more understandable.

In the following, this section will be split in two subsections: Structural Section of TDevC Sect. 6.5.1.1 and Behavioral Section of TDevC Sect. 6.5.1.2.

6.5.1.1 Structural Description of a TDevC Device Model

The structural section of the TDevC device model is composed by the specification of device registers, registers formats, and data patterns through the constructors *register*, *format* and *pattern*. Listing 16 shows an example of these constructions.

```
device (dm9000a) {
    pattern RXNOERROR = mask(....0000);

    format physicalAddrfmt {
        RW PAB[7:0];
    }

    external register indexReg(0x00) alias
    INDEXREG {
         RW INDEX [15:0];
    }

```

```
13     internal IntRegsProt register
14     networkStatusReg(0x00) alias NSR{
15        READ SPEED [7];
16        READ LINKST [6];
17        RW WAKEST [5];
18        reserved [4];
19        RW TX2END [3];
20        RW TX1END [2];
21        READ RXOV [1];
22        reserved [0];
23     }
24
25     internal IntRegsProt register phyAddrReg5(0x15) alias
           PAR5 = physicalAddrfmt;
26  ...
```

Listing 16 Example of the structural section of TDevC device model

First of all, every TDevC specification starts with the constructor *device (device_name)*, as shown in Listing 16-line 1, where *device_name* is the name of the device under validation.

In the following, Listing 16-line 2 shows an example of a *pattern* declaration. Using *pattern* is the way to specify numbers and mask patterns. It makes the behavior description cleaner and less error-prone. It looks like a constant variable of general purpose languages, but, besides a fixed value, it can also represent a fixed format of data. In line 3 of Listing 16 there is the declaration of *RXNOERROR* pattern, which represents the absent of error during the package transmission. In this example, the *mask* constructor means that any register field or variable compared to this pattern must contain the last four bit equal to zero to be true.

The *register* constructors, as the name already say, reflect the physical registers of the device. They can be explicitly declared with its visibility scope, name, alias, fields, physical addresses, and access permissions, as shown in Listing 16-lines 8 to 21, or can use a previous format that was declared using the *format* constructor. The *format* constructor is declared similarly to the *register* constructor. They are used to define a format common to various registers. By this way, the designer must declare the format once, as shown in Listing 16-lines 4 to 6, and then bind the register to its respective format, adding the register's physical address and alias, as it is shown in Listing 16-lines 23.

As it can be seen in the Listing 16-lines 8 and 12, there are two types of visibility scope for registers: external, represented by the construction *external* and internal, represented by the construction *internal*. External registers are those that are mapped to platform addressing range, i.e., the registers are directly accessed by the platform's master elements. On the other hand, internal registers are those that are not addressed directly in the range of system address. They are accessed through an access protocol, commonly implemented by the hardware-dependent software layer, and uses the external registers as input. Listing 16-lines 24 and 22 shows examples of two internal registers using an access protocol called *IntRegsProt*. The declaration and construction of the protocols will be detailed in Sect. 6.5.1.2.

It is important to highlight that the registers addressing is absolute within the range of devices, however, to external registers, it is about the platform's addressing.

In the case of the external register *indexReg*, it can be seen that its address in the device scope is *0x00*. However, if the device in question is in the system address range *0x00A-0x00E*, its relative address in the device's addressing remains *0x00*, where it is translated to the absolute address *0x00A* on the platform's addressing.

Also, concerning registers addressing, external registers are in a different address scope than internal registers, since the external register has its addresses linked and translated directly to the target platform and the internals registers are dependent on the access protocols.

So with this relationship between the internal registers and its protocols, it becomes evident that each defined protocol carries with it a different scope of addressing, allowing internal registers with different protocols and external registers sharing the same address numerical value. An example of this can be seen in the Listing 16-line 8 and 12.

The optional attribute *alias*, as the name already says, defines an alias to the register and can be used at behavioral sections of the model. The purpose of this constructor is to allow the simplified registers reference in the specification and at the same time to maintain its complete and precise description. Typically the *datasheets* includes the registers by their full name and refer an alias, usually their initials, in their applications.

Fields of registers, as it can be seen in Listing 16-lines 5, 9, and 13 to 20, are logical subdivisions, commonly described in the *datashrets* of devices. Typically each subdivision has a specific function. A value assigned to a field can lead to a behavior of the device. Therefore, to clarify the description in TDevC and reduce the possibility of errors in the comparison of registers values, the language supports specifying nested fields. Registers contain fields, their fields, consequently, may also contain fields and so on.

Additionally, another mandatory attribute is the field name. Every field must contain a name, even if it has subfields. These names are used in the behavioral section of TDevC. Fields of a register are referenced by registers' name or *alias* followed by the field, only separated by a ".". The same pattern can be used for fields, subfields, and so on. An example of this constructor is showed in the following.

```
indexReg.INDEX
```

If, for example, the field *INDEX* of the register *indexReg* had a subfield called *INDEXHI*, its reference would be made like the following TDevC code.

```
indexReg.INDEX.INDEXHI
```

and so on.

6.5.1.2 Behavioral Specification of a TDevC Device Model

The behavioral specification of a TDevC device model consists of constructors with the syntax for protocol declaration, context variables, and the HFSM-D, its data assignment and assertions. For the sake of a better understanding, this document will address in the following, each one of these constructors, starting with the protocols declarations, following with the declarations of variables and the HFSM-D state machines, ending with the declaration of assertions.

```
1  protocol IntRegsProt {
2      address: INDEXREG(0X00);
3      data: DATAREG;
4  }
5
6  protocol ProtPHYRegs{
7      address: EPAR.REGADDR(0X00);
8      data: {EPDR.EE_PHY_H;EPDR.EE_PHY_L}
9      readingtrigger{
10         write(EPCR) = 0x0C;
11     }
12     writingtrigger{
13         write(EPCR) = 0x0A;
14     }
15 }
```

Listing 17 Example of protocol declaration in behavioral section of a TDevC device model

As explained in the previous section, protocols are used to define the access procedures of internal registers, from access to external recorders, directly or indirectly. The Listing 17 shows two examples of this construction.

The protocol block includes the reserved word *protocol* followed by an identifier. Within the block the designer can specify the registers or register fields representing the address and the data of the internal register covered by that protocol through the constructors *address* and *data*, respectively. Listing 17-lines 2, 3, 7, and 8 shows examples of these constructors.

In Listing 17-lines 2 and 3, the protocol *IntRegsProt* describes that all internal registers covered by it will be accessed whenever their address is in the register *INDEXREG* and there is a reading or writing of a value in the register *DATAREG*. The same can be observed in the Listing 17-lines 7 and 8. However, the register field that defines the address of all registers covered by the protocol *ProtPHYRegs* is the *EPAR.REGADDR* and the register fields that define the data read or written from registers covered by this protocol are *EPDR.EE_PHY _H* and *EPDR.EE_PHY_L*. In the protocol *ProtPHYRegs* the attribute data is formed by concatenated fields, once *EPDR.EE_PHY_H* is the most significant part and *EPDR.EE_PHY_L*, the less significant.

In some cases, it is necessary to inform a trigger that defines when the data is ready to be read or written. For this, the constructors *readingtrigger* and *writingtrigger* have been defined. The specification of Ethernet DM9000A says that to perform readings and writings to internal registers thought this protocol it is necessary to write the values *0x0C* and *0x0A*, respectively, in register *EPCR* informing that the address and the data are configured. This feature is captured in Listing 17-lines 9–14.

It is important to clarify that the use of the constructor *protocol* aims to simplify the specification, making it clear and direct the visualization when the registers are being used. However, one can describe a protocol in TDevC defining the access to internal registers indirectly using external register only on all properties and transactions of the state machine, which would make the specification rather obscure, but correct.

```
1  var t1 = 1;
2  var rxlowlen;
3  var pkgcounter;
4  var rxlen;
5
6  globalstate {
7      orthoregion ethOperationMode {
8          initialstate UNDEF_OPER_MODE {
9              addexitpoint(OPER16BITS){
10                 read(ISR.IOMODE) == 0
11             }
12
13             addexitpoint(OPER8BITS){
14                 read(ISR.IOMODE) == 1
15             }
16
17             addentrypoint{
18                 write(CR.RST) == 1
19             }
20         }
21         state OPER16BITS {
22             addexitpoint(OPER8BITS){
23                 read(ISR.IOMODE) == 1
24             }
25         }
26         state OPER8BITS {
27             addexitpoint(OPER16BITS){
28                 read(ISR.IOMODE) == 0
29             }
30         }
31     }
32 ...
```

Listing 18 Example of Variables and HFSM-D declarations in behavioral section of a TDevC device model

The declaration of variables and the HFSM-D state machine, as it can be seen in the example in Listing 18, uses the constructors *var* and *globalstate* respectively.

The variables are used when you want to add some context for validation. The use of variables, its assignments, and the HFSM-D state machine makes all the validation *stateful*. These constructors were supported in the previous versions of the TDevC language, where validation had no context and no notion of execution states, being completely *stateless*. In the current version, every variable has a global scope and all assignments occur during transitions of states in the HFSM-D state machine. The Listing 18-lines 1 to 4 show examples of variable declarations. In Listing 18-line 1 it is shown exactly a variable declaration with assignment of an initial value.

Since all variables were declared, now it can specify the HFSM-D state machine. Indeed, even the reserved word used already makes clear, *globalstate* indicates the root state, unique, and parent state of all state machine's states, on the first hierarchical level of the HFSM-D state machine. As well as any state of the hierarchical state machine, the global state is segmented into orthogonal regions through subblocks starting with the reserved word *orthoregion*, respecting the definition of the HFSM-D state machine described in Sect. 6.4. The Listing 18-line 6 shows the beginning of the global state block and, hence, of the HFSM-D state machine.

Each orthogonal region is an independent execution line within a state. Thus, its syntax also contains the declaration of child states, allowing the declaration of a hierarchical machine, as it can be seen in Listing 18 lines 8, 21, and 26.

Every orthogonal region must have a single initial state declared by the constructor initiated by the reserved word *initialstate*. Other states are optional, however, according to the definition of hierarchical machine, all other states are intermediate, declared by the constructor initiated by the reserved word *state*.

The example shown in Listing 18 contains an orthogonal region called *ethOperationMode*, located in line 7. Its initial state, in line 8, is the *UNDEF_OPER_MODE* and their two intermediate states, located in lines 21 and 26 respectively, are called *OPER16BITS* and *OPER8BITS*. This orthogonal region is the selection of the operating mode of the Ethernet DM9000A. It is initially undefined for the embedded software layer and, according to the selection, becomes 8-bit or 16-bit.

Inside the state definition blocks, it can be seen in Listing 18-lines 9, 13, 22, and 27 the use of the attribute *addexitpoint*, and in line 17 the use of the attribute *addentrypoint*. These constructors are used to specify the state transitions. As mentioned in the previous section, there are two types of transitions: *exit points* and *entry points*. Thus, these two constructors are used, respectively, for specifying them.

Taking the exit point specified in Listing 18-line 9, it can be seen a transition from the state *UNDEF_OPER_MODE* to the state *OPER16BITS*, depending if the Boolean expression *read(ISR.IOMODE) == 0* is true, that specified inside the block. This expression is true when the embedded software layer knows, through a reading access, that the register field *ISR.IOMODE* contains the value zero (0). The attribute *addentrypoint* works the same way; however, it does not require a target state.

```
1  addproperty(critical) UndefinedOperMode{
2      ltlf([](!UNDEF_OPER_MODE))
3  }
4  addproperty(critical) WriteBeforeLen{
5      ltlf(!TXWRPKGLEN U TXWRPKGWR)
6  }
7  addproperty(critical) LenBeforeSend{
8      ltlf(!TXSDPKGSDING U TXWRPKGLEN)
9  }
10 addproperty(warning) NeverLenAndSend {
11     ltlf([]( !(TXSDPKGSDING && TXWRPKGLEN)))
12 }
```

Listing 19 Example of assertions declarations in behavioral section of a TDevC device model

Another important attribute specified inside the state block is the *addproperty*. It is used to define the assertions a particular state must check. Listing 19 shows examples of assertions specified inside the state *PHYUP*. The state *PHYUP* means that the physical layer of the DM9000A is on and it is ready to transmit or to receive network packages.

As mentioned earlier, the assertions are specified through an LTL formulae, expressed through the attribute *ltlf*, as it can be seen in Listing 19-lines 2, 5, 8, and 11. The LTL formulae show the properties that must be held. So, for example, using the assertion in Listing 19-lines 2, the LTL formulae is equivalent to $G(\sim UNDEF\_OPER\_MODE)$, what means that always the state

UNDEF_OPER_MODE must be negated while inside the state *PHYUP*. In other words, the physical layer of the DM9000A cannot be on without knowing its operation mode.

The attribute *addproperty* also supports specifying a qualitative information, that can be: *critical*, *warning* and *info*. In the current version of TDevC, only the type *critical* has a different treatment. It means that such assertion is critical for the whole system, and it may lead to a system execution for a critical and unpredictable state, and maybe generating a desynchronization of the HFSM-D state machine. Facing a critical assertion, the MDDC signalizes the event and stops the validation until a system reset. The others types of assertions are, for now, only informative and they can be used for modeling the TDevC platform model. With these properties, a meta-model, called TDevC platform model, can join the properties feedback from the TDevC device models to perform a complete validation, involving the whole platform. The TDevC platform model will be detailed in the next section.

6.5.2 *TDevC Platform Model*

Differently from the TDevC device models, the platform models do not contain structural section. Their primary goal is modeling the behaviors and properties of the whole platform, only based on the devices' behaviors. So for the meta-models, the devices' structures are completely transparent and irrelevant.

So, the main difference between the two models is that before the beginning of the HFSM-D state machine declaration, in the platform model the designer must specify which devices will be used for that meta-model.

For the syntax presentation in this section, an example of two separated simple devices will be used, which control a bank of sensors and a bank of actuators. These devices contain, each one, four 1-bit sized ports called channels. The sensors device receives from the input channels the activation signals of the physical sensors and the actuators device sends to the output channels the activation signal to the physical actuators. However, the embedded software will decide and control when to activate an actuator depending on a feedback from the sensors. For this example, the sensors channels are numerically correspondent to the actuators channels, i.e., the sensor channel 1 is used with the actuator channel 1 and so on, for all channels.

```
import "mysensors.tdevc";
import "myactuators.tdevc";

platform (uClinuxNiosV2) {

     device mysensors alias mys;
     device myactuators alias mya;
...
```

Listing 20 Example of declaration of a TDevC platform model

The first constructor that must be taken into account is the *import*. Through this constructor, the meta-model defines that devices models will be used. Just to make clear, the device models' assertions are still fired by the MDDC regardless of platform model. Listing 20-lines 1 and 2 show examples of the constructor *import*.

The meta-model definitions start with the reserved word *platform*, followed by an identifier, in which, for the example of Listing 20-line 4, is the *uClinuxNiosV2*. This identifier is necessary to distinguish different platform models covering different properties for the same hardware platform. This feature is interesting once several systems and applications can be set for the same platform configuration. By using a fixed hardware set, it is possible to vary the whole software layer and the system application. Thus, it is possible to have a platform models library composed by models covering different system's properties, only depending on the layer software configuration. The platform design varies according to the system purpose, unlike the device models, that are fixed for each device.

In the following, the designer declares all devices that will be used in HFSM-D state machine. The constructor *device* is used to define an alias for each device. The alias, as the registers alias in the device models, simplifies the code making the codification agiler. This constructors can be seen in Listing 20-lines 6 and 7.

```
1  var chanLen = 4;
2  globalstate {
3      addproperty(critical) Channel1Safety {
4          ltlf(
5              [](mys.Channel1On -> mya.Channel1On)
6              )
7      }
8      addproperty(critical) Channel2Safety {
9          ltlf(
10             [](mys.Channel2On -> mya.Channel2On)
11             )
12     }
13     addproperty(critical) Channel3Safety {
14         ltlf(
15             [](mys.Channel3On -> mya.Channel3On)
16             )
17     }
18     addproperty(critical) Channel4Safety {
19         ltlf(
20             [](mys.Channel4On -> mya.Channel4On)
21             )
22     }
23   }
```

Listing 21 Example of assertions declarations in a TDevC platform model

The last syntactical portion of the platform model includes the variable declaration and the HFSM-D state machine definition, already explained in the previous sections. There is not any syntactical difference, only semantic variations.

This semantic difference is related to the atomic propositions of the state transitions' Boolean expression and the LTL formulae. Once the platform models do not have any structural component, replacing the registers and registers fields references, they use assertions' feedback from the instantiated devices.

As it can be seen in Listing 21-lines 5, 10, 15 and 20, the atomic propositions used in the LTL formulae are composed by the device, followed by the identifier of the assertion, separated only by a ".". For example, in line 5, the LTL expression says that whenever (operator *always*) the device assertion *Channel1On* from the device *mysensors* is held, the assertion *Channel1On* from the device *myactuators* must be held.

6.6 Architecture of the Monitoring Module

As said before, from a TDevC specification the monitoring module called MDDC is synthesized. As it is shown in Fig. 6.7, the monitor is composed of four main components: Bus Slave Interface (BSI), Bus Snooping Interface (BSPI), Protocol Translator (PT) and FSM Controller. BSI is the interface that connects the monitor to the platform as a standard slave peripheral or an external host. Through this interface, the processing element or a remote host knows when a property is signalized, and it can request information about the validation, such as last accessed devices, states, and behavioral properties.

The BSPI is the gateway of the validation. Through this interface all the accesses made by the processing element are captured, identified and, if they belong to an under validation device, they are dispatched to the component PT (protocol translator).

The protocol translator, as the name implies, translates the protocols used to access internal registers of the devices. Based on the protocols specified in the high-level models, the tool-set synthesizes the PT. The protocol information is used to infer which internal register the embedded software is accessing. Thus, it is possible to perform the state transitions based on internal registers.

After the protocol translation, accesses' information is sent to the finite-state machine (FSM) Controller. This component contains all the state machines of all models located into the subcomponent FSM-Set.

There are two types of FSM modules into the FSM-Set: FSM-Devices, for each device under validation, and one FSM-Platform, representing the platform meta-model, as it can be seen in Fig. 6.7. All FSM-Devices are connected to the FSM-

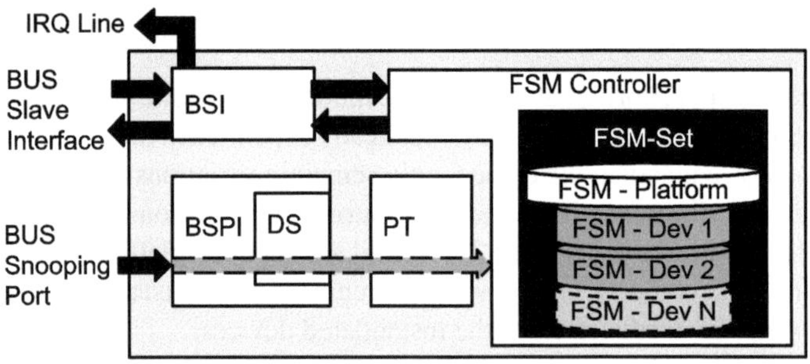

Fig. 6.7 Architecture of the monitoring module

platform through feedback ports. These feedback ports are output ports of the FSM-devices that signalizes the occurrence of assertions.

The FSM Controller can request information about each state machine and sends them to the BSI. Every time an assertion is fired the FSM Controller informs the BSI about it. Thus, the BSI uses the IRQ Line to inform the processing element about the occurrence of the event and, depending on the need, the processing element can request that information through the slave interface.

6.7 Experiments and Results

For validating the effectiveness of the approach, some experiments were done using four devices: a DM9000A Ethernet controller, a sensors bank controller, and a fan controller, composing a temperature control system, and an Altera UART controller. All experiment have used a platform based on the NIOS II processor and running a embedded μClinux. Figure 6.8 shows the infrastructure used for the experiments.

The experiments performed with the DM9000A Ethernet Device controller covered the physical layer (PHY) services, such as power down, power up, PHY reset, data transmission, and reception, the definition of the operation mode and state link monitoring. These experiments were performed regarding only the DM9000A. Thus, There was not any reference to the DM9000A device in the TDevC platform model.

The experiment were done using the UNIX tool *ifconfig* to start and shut down the Ethernet's physical layer and the *SSH* protocol and a web server to access a remote machine and perform packages transmission and reception. The TDevC device model was specified in the proposed DSL, and the DM9000A device driver was modified to add randomly unwanted behaviors during its execution. With no impact on the

Fig. 6.8 Real infrastructure used for the experiments

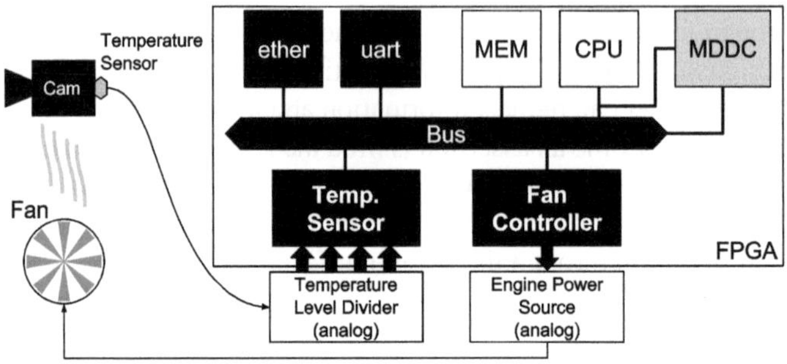

Fig. 6.9 Temperature control system's diagram

execution time, the technique detected all purposeful failures, instantly when they occur. All feedback of the behavior detection was made through LEDs installed on the hardware board. Thus, there was not any software interference, including CPU interruption.

The temperature control system was fully built to validate the proposed approach. The system is composed of two separate devices, as said before. As it can be seen in Fig. 6.9, the temperature sensor consists of an analog temperature sensor, with its signal being converted to four 1-bit size channels (four temperature zones) through an analog temperature level divider. The fan controller is actuator used to reduce the temperature of the measured object. The sensors and the actuators are independents and are software's responsibility to use and control them.

The platform used in this experiments remains the same of other experiments, but now running a particular application, also built for this approach.

As it is shown in Fig. 6.9, the sensor measures the temperature of a night vision camera, which heats when the air conditioning is off and at night, because of the infrared lights. Thus, the primary constraint on the whole application is to guarantee that the temperature stays always below a certain threshold. The fan cannot turn off before the temperature reaches a safe zone and it should turn off after the temperature reaches a safe zone.

This experiment aims to validate if the approach can identify assertions regardless of their location in the software stack. So, For this purpose, the TDevC models were specified and synthesized. Due to the purpose of these experiments and the number of devices involved, a TDevC platform model had to be created.

Two procedures have been adopted for these experiments: Shut down the application and to insert a failure in the application, turning on and off the fan randomly.

For the two procedures, both device and platform assertions were instantly detected. Again, all feedback of the behavior detection was made through LEDs installed on the hardware board, what means that there was not any software interference, including CPU interruption. Like the DM9000A experiment, there was not any overhead in the execution time.

For the Altera UART, it was developed a driver used to control the MDDC and a monitoring application used to control it from the command console. The monitoring module's driver was written to provide management capability over its registers. Also, it is responsible for treating the interruptions generated by errors found by the module.

A Linux application that uses this driver was written to provide the user with a simple way to interact with the monitoring module, turning it on or off and reading information about previous validation errors detected.

From the UART datasheet, two properties were extracted to be verified by the validation mechanism. These properties are related to the requirements for reading and writing data from the received data buffer and the transmitting data buffer. Such as the DM9000A, these experiments were also performed regarding only the Altera UART. Thus, There was not any reference to the UART device in the TDevC platform model.

For testing the impact of the monitoring module on the system performance, two tests were made with the UART. One of the tests no errors were inserted in the UART driver, on the other, errors that violated the previously commented properties extracted from the UART datasheet were inserted. The time to transfer a sequence of 100 kilobytes through the UART interface was measured in each test.

These experiments have showed that, when there were no errors in the UART driver, the presence of the monitoring module had no effect on the performance, in fact, the time was even better with it. But this difference in times is due only to the nature of the operating system and it is context switching that makes the execution times of the experiment varies each time it is executed.

When errors were inserted in the driver, the use of the Monitoring Module represented a loss of about 1% in performance. This feature is due to the Monitoring Module generates interruptions each time a property violation is detected. These interruptions make the processor stop its current activity to treat them. During the tests, all properties violations inserted in the driver were detected and reported correctly.

Concluding the experiments, Table 6.1 shows the added overhead during the platform execution, for each device. Besides, the table summarizes if there was any software interference.

Continuing with the experiments results, the Table 6.2 presents, for each device, the number of code lines of the model, using the high-level DSL, and the number o code lines that the toolset has generated after the synthesis.

Table 6.1 Additional runtime overhead and software interference

Device	Additional overhead (%)	Interference
Ethernet	0	No
Temp. controller	0	No
UART	1	Yes

Table 6.2 Developed code sizes (Line of codes)

Device	Device model	Platform model	Synthesized monitor
Ethernet	623	0	1682
Temp. controller	134	34	1639
UART	96	0	705

Table 6.3 FPGA area usage (Logic elements)

Device	Platform	Isolated monitor	%
Ethernet	7620	870	11.4
Temp. controller	7620	864	11.3
UART	4576	390	8.5

Finally, Table 6.3 shows the FPGA area usage, for each device, covering the whole platform, the single monitor, and the relation between the monitor area and the platform area.

6.8 Conclusions

The experiments have shown that the use of this technique has increased the reliability of systems, detecting unwanted behaviors, and increasing the resilience, predictability, and adaptivity of systems by reporting the occurrence of specified behaviors.

They also have shown the importance of this technique applied for CPS, finding improper misunderstanding of sensors feedback or usage of actuators, preventing, for example, the degradation of components or, the must important, loss of life.

Another significant improvement in system design is the low time consumption used for modeling and performing the system validation and debugging. Once this technique proposes a synthesis of high-level models, the time would be used to develop the whole validation environment is significantly reduced.

6.8.1 Future Works

The next main feature is to support multicore platforms. Introducing several processing elements in a single platform can cause serious concurrency problems. Sharing almost the same resources, they must respect critical sessions, synchronizing their accesses. To guarantee that the sharing is being performed correctly, the validation environment must know the critical sections and must be able to identify behavioral issues such as deadlock and starvation.

The first stage of the design of this new feature, in which is already in progress, is the TDevC models' adaptation. Each state of the HFSM-D or a group of them can be associated with a critical section. Thus, for example, if two or more processing elements lead, at the same moment, the device execution to the same critical section, the monitor must signalize a violation of that critical section.

Acknowledgements This research is being supported by Brazilian Research Council CNPq (Grant nr. 485829/2012-6).

References

1. Acquaviva A, Bombieri N, Fummi F, Vinco S (2013) Semi-automatic generation of device drivers for rapid embedded platform development. IEEE Trans CAD Integr Circuits Syst 32(9):1293–1306
2. Amani S, Chubb P, Donaldson A, Legg A, Ong KC, Ryzhyk L, Zhu Y (2014) Automatic verification of active device drivers. ACM Operating Syst Rev 48(1)
3. Baheti R, Gill H (2011) Cyber-physical systems. The impact of control technology, vol 12, pp 161–166
4. Baier C, Katoen J-P (2008) Principles of model checking (Representation and mind series). The MIT Press
5. Behrend J, Gruenhage A, Schroeder D, Lettnin D, Ruf J, Kropf T, Rosenstiel W (2014) Optimized hybrid verification of embedded software. In: Test workshop—LATW, 2014 15th Latin American, pp 1–6, March 2014
6. Behrend J, Lettnin D, Heckeler P, Ruf J, Kropf T, Rosenstiel W (2011) Scalable hybrid verification for embedded software. In: DATE, pp 179–184. IEEE
7. Ganapathy V, Balakrishnan A, Swift MM, Jha S (2007) Microdrivers: a new architecture for device drivers. In: Hunt GC (ed) HotOS. USENIX Association
8. Gastin P, Oddoux D (2001) Fast ltl to büchi automata translation. In: Berry G, Comon H, Finkel A (eds) Computer aided verification. Lecture notes in computer science, vol 2102. Springer, Berlin, pp 53–65
9. Harel D (1987) Statecharts: a visual formalism for complex systems. Sci Comput Program 8(3):231–274
10. Herder JN, Bos H, Gras B, Homburg P, Tanenbaum AS (2007) Failure resilience for device drivers. In: Proceedings of the 37th annual IEEE/IFIP international conference on dependable systems and networks, DSN '07, pp 41–50, Washington, DC, USA. IEEE Computer Society
11. Kadav A, Renzelmann MJ, Swift MM (2009) Tolerating hardware device failures in software. In: Proceedings of the ACM SIGOPS 22nd symposium on operating systems principles, SOSP '09, pp 59–72, New York, NY, USA. ACM
12. Lee EA (2008) Cyber physical systems: design challenges. Technical Report UCB/EECS-2008-8, EECS Department, University of California, Berkeley, Jan 2008
13. Lettnin D, Nalla PK, Behrend J, Ruf J, Gerlach J, Kropf T, Rosenstiel W, Schonknecht V, Reitemeyer S (2009) Semiformal verification of temporal properties in automotive hardware dependent software. In: Design, automation test in Europe conference exhibition, 2009. DATE '09, pp 1214–1217, April 2009
14. Lisboa E, Silva L, Chaves I, Lima T, Barros E (2009) A design flow based on a domain specific language to concurrent development of device drivers and device controller simulation models. In: Proceedings of the 12th international workshop on software and compilers for embedded systems, SCOPES '09, pp 53–60, New York, NY, USA. ACM
15. Macieira RM, Barros E, Ascendina C (2014) Towards more reliable embedded systems through a mechanism for monitoring driver devices communication. In: 2014 15th international symposium on quality electronic design (ISQED), pp 420–427, March 2014

16. Macieira RM, Lisboa EB, Barros ENS (2011) Device driver generation and checking approach. In: 2011 Brazilian symposium on computing system engineering (SBESC), pp 72–77, Nov 2011
17. Reinbacher T, Brauer J, Horauer M, Steininger A, Kowalewski S (2014) Runtime verification of microcontroller binary code. Sci Comput Program 80, Part A(0):109–129. Special section on foundations of coordination languages and software architectures (selected papers from FOCLASA'10), Special section—Brazilian Symposium on Programming Languages (SBLP 2010) and Special section on formal methods for industrial critical systems (Selected papers from FMICS'11)
18. Ryzhyk L, Chubb P, Kuz I, Le Sueur E, Heiser G (2009) Automatic device driver synthesis with termite. In: Proceedings of the ACM SIGOPS 22nd symposium on operating systems principles, SOSP '09, pp 73–86, New York, NY, USA. ACM
19. Sistla AP, Clarke EM (1985) The complexity of propositional linear temporal logics. J ACM 32(3):733–749
20. Swift MM, Martin S, Levy HM, Eggers SJ (2002) Nooks: an architecture for reliable device drivers. In: Proceedings of the 10th workshop on ACM SIGOPS European workshop, EW 10, pp 102–107, New York, NY, USA. ACM
21. Villarraga C, Schmidt B, Bao B, Raman R, Bartsch C, Fehmel T, Stoffel D, Kunz W (2014) Software in a hardware view: new models for hw-dependent software in soc verification and test. In: 2014 IEEE international test conference (ITC), pp 1–9, Oct 2014
22. Weggerle A, Himpel C, Schmitt T, Schulthess P (2011) Transaction based device driver development. In: MIPRO, pp 195–199. IEEE

Chapter 7
Model Checking Embedded C Software Using k-Induction and Invariants

Herbert Rocha, Hussama Ismail, Lucas Cordeiro and Raimundo Barreto

7.1 Introduction

The Bounded Model Checking (BMC) techniques based on Boolean Satisfiability (SAT) [8] or Satisfiability Modulo Theories (SMT) [2] have been successfully applied to verify single- and multi-threaded programs and to find subtle bugs in real programs [11, 12, 25]. The idea behind the BMC techniques is to check the negation of a given property at a given depth, i.e., given a transition system M, a property ϕ, and a limit of iterations k, BMC unfolds the system k times and converts it into a Verification Condition (VC) ψ such that ψ is *satisfiable* if and only if ϕ has a counterexample of depth less than or equal to k.

Typically, BMC techniques are only able to falsify properties up to a given depth k; however, they are not able to prove the correctness of the system, unless an upper bound of k is known, i.e., a bound that unfolds all loops and recursive functions to their maximum possible depth. In particular, BMC techniques limit the visited regions of data structures (e.g., arrays) and the number of loop iterations. This limits the state space that needs to be explored during verification, leaving enough that real errors in applications [11, 12, 21, 25] can be found; BMC tools are, however, susceptible to exhaustion of time or memory limits for programs with loops whose bounds are too large or cannot be determined statically.

Consider, for example, the simple program in Listing 22 (left), in which the loop in line 2 runs an unknown number of times, depending on the initial nondeterministic value assigned to x in line 1. The assertion in line 3 holds independent of x's initial

H. Rocha (✉)
Federal University of Roraima, Boa Vista, Brazil
e-mail: herberthb12@gmail.com

H. Ismail · L. Cordeiro · R. Barreto
Federal University of Amazonas, Manaus, Brazil
e-mail: hussamaismail@gmail.com

L. Cordeiro
e-mail: lucasccordeiro@gmail.com

R. Barreto
e-mail: xbarretox@gmail.com

D. Lettnin and M. Winterholer (eds.), *Embedded Software Verification and Debugging*, Embedded Systems, DOI 10.1007/978-1-4614-2266-2_7

value. Unfortunately, BMC tools like CBMC [11], LLBMC [25], or ESBMC [12] typically fail to verify programs that contain such loops. Soundness requires that they insert a so-called *unwinding assertion* (the negated loop bound) at the end of the loop, as in Listing 23 (right), line 5. This *unwinding assertion* causes the BMC tool to fail if k is too small.

```
unsigned int x=*;
while(x>0) x—;
assert(x==0);
```

Listing 22 Unbounded loop

```
1 unsigned int x=*;
2 if (x>0)     } k copies
3   x—;        }
4 ...
5 assert(!(x>0));
6 assert(x==0);
```

Listing 23 Finite unwinding

In mathematics, one usually attacks such unbounded problems using *proof by induction*. A variant called k-induction has been successfully combined with continuously refined invariants [6], to prove that (restricted) C programs do not contain data races [14, 15], or that design-time time constraints are respected [16]. Additionally, k-induction is a well-established technique in hardware verification, where it is easy to apply due to the monolithic transition relation present in hardware designs [16, 18, 32]. This paper contributes a new algorithm to prove correctness of C programs by k-induction in a completely automatic way (i.e., the user does not need to provide the loop invariant).

The main idea of the algorithm is to use an iterative deepening approach and check, for each step k up to a maximum value, three different cases called here as base case, forward condition, and inductive step. Intuitively, in the base case, we intend to find a counterexample of ϕ with up to k iterations of the loop. The forward condition checks whether loops have been fully unrolled and the validity of the property ϕ in all states reachable within k iterations. The inductive step verifies that if ϕ is valid for k iterations, then ϕ will also be valid for the next unfolding of the system. For each step of the algorithm, we infer program invariants using affine constraints to prune the state space exploration and to strengthen the induction hypothesis.

These algorithms were all implemented in the Efficient SMT-based Context-Bounded Model Checker tool (known as ESBMC[1]), which uses BMC techniques and SMT solvers (e.g., [10, 13]) to verify embedded systems written in C/C++ [12]. In Cordeiro et al. [12] the ESBMC tool is presented, which describes how the input program is encoded in SMT; what the strategies for unrolling loops are; what are the transformations/optimizations that are important for performance; what are the

[1] Available at http://esbmc.org/.

benefits of using an SMT solver instead of a SAT solver; and how counterexamples to falsify properties are reconstructed. Here we extend our previous work in Rocha et al. [29] and Ramalho et al. [17] and focus our contribution on the combination of the k-induction algorithm with invariants. First, we describe the details of an accurate translation that extends ESBMC to prove the correctness of a given (safety) property for any depth without manual annotations of loops invariants. Second, we adopt program invariants (using polyhedra) in the k-induction algorithm, to speed up the verification time and to improve the quality of the results by solving more verification tasks in less time. Third, we show that our present implementation is applicable to a broader range of verification tasks, which other existing approaches are unable to support [14, 15, 18].

7.2 Motivating Example

As a motivating example, a program extracted from the benchmarks of the SV-COMP [3] is used as a running example as shown in Listing 24, which already includes invariants using polyhedra.

```
1  int main(int argc, char **argv)
2  {
3   uint64_t i=1, sn = 0;
4   assume( i==1 && sn==0 ); // Invariant
5   uint32_t n;
6   assume(n>=1);
7   while (i<=n) {
8     assume( 1<=i && i<=n ); // Invariant
9     sn = sn+a;
10    i++;
11  }
12  assume( 1<=i && n+1<=i ); // Invariant
13  assert(sn==n*a);
14 }
```

Listing 24 Running example for the k-induction algorithm.

In Listing 24, a is an integer constant and note that variables i and sn are declared with a type larger than the type of the variable n to avoid arithmetic overflow. Mathematically, the code above represents the implementation of the sum given by the following equation:

$$S_n = \sum_{i=1}^{n} a = na, n \geq 1 \tag{7.1}$$

In the code of Listing 24, the invariants produced by PIPS are included as assume statements; the property (represented by the assertion in line 13) must be *true* for any value of n (i.e., for any unfolding of the program). In contrast from our k-induction algorithm, BMC techniques have difficulties in proving the correctness of this (simple) program since the upper limit value of the loop, represented by n, is nondeterministically chosen (i.e., the variable n can assume any value from one to the size of the *unsigned int* type, which varies between different types of computers). Due to this condition, the loop will be unfolded $2^n - 1$ times (in the worst case, $2^{32} - 1$ times on 32 bits integer), which is thus impractical. Basically, the bounded model checker would symbolically execute several times the increment of the variable i and the computation of the variable sn by $4, 294, 967, 295$ times. To solve the problem of unfolding the loop $2^n - 1$ times, the translations previously described are performed.

7.3 Induction-Based Verification of C Programs Using Invariants

The transformations in each step of the k-induction algorithm take place at the intermediate representation level, after converting the C program into a GOTO-program, which simplifies the representation and handles the unrolling of the loops and the elimination of recursive functions.

7.3.1 *The Proposed k-Induction Algorithm*

Listing 25 shows an overview of the proposed k-induction algorithm. We do not add additional details about the transformations in each step of the algorithm; we keep it simple and describe the details in the next subsections so that one can have a big picture of the proposed method. The input of the algorithm is a C program P together with the safety property ϕ. The algorithm returns *true* (if there is no path that violates the safety property), *false* (if there exists a path that violates the safety property), and *unknown* (if it does not succeed in computing an answer *true* or *false*).

In the base case, the algorithm tries to find a counterexample up to a maximum number of iterations k. In the forward condition, global correctness of the loop with regard to the property is shown for the case that the loop iterates at most k times; and in the inductive step, the algorithm checks that, if the property is valid in k iterations, then it must be valid for the next iterations. The algorithm runs up to a maximum number of iterations and only increases the value of k if it cannot falsify the property during the base case.

7.3.1.1 The Difference to Other k-Induction Algorithms

Our k-induction algorithm is slightly different than those presented by Große et al. [18], Donaldson et al. [15], and Hagen et al. [19]. In Große et al., the forward condition and the inductive step are computed differently from our approach (as described in Sect. 7.3.1) and the value of k is increased only at the end of the algorithm; in this particular case, computational resources are thus wasted since loops are usually unfolded at least two times. Donaldson et al. [15] and Hagen et al. [19] propose the k-induction with two steps only (i.e., the base case and the inductive step); however, the inductive step of the approach proposed by Donaldson et al. requires annotations in the code to introduce loops invariants. It is worth noting that Donaldson et al. improve the method and reduce the annotation overhead [14]. However, our method is completely automatic as in Hagen et al. [19]. Additionally, as observed in the experimental evaluation (see Sect. 7.4), the use of the forward condition, in our proposed method, improves significantly the quality of the results, because some programs that are hard to be proved by the inductive step can be proved by the forward condition using affine constraints.

```
input: program P and safety property ϕ
output: true, false, or unknown
k = 1
while k <= max_iterations do
  if base_case(P, ϕ, k) then
     show counterexample s[0..k]
    return false
  else
    k=k+1
    if forward_condition(P, ϕ, k) then
      return true
    else
      if inductive_step(P, ϕ, k) then
        return true
      end-if
    end-if
  end-if
end-while
return unknown
```

Listing 25 An overview of the k-induction algorithm.

7.3.1.2 Loop-Free Programs

In the k-induction algorithm, the loop unwinding of the program is done incrementally from one to *max_iterations* (cf. Listing 25), where the number of unwindings is measured by counting the number of *backjumps* [27]. In each step of the k-induction algorithm, an instance of the program that contains k copies of the loop body corresponds to checking a loop-free program, which uses only *if*-statements in order to prevent its execution in the case that the loop ends before k iterations.

Definition 7.1 (*Loop-free Program*) A loop-free program is represented by a straight-line program (without loops) by providing an $ite\ (\theta, \rho_1, \rho_2)$ operator, which takes a Boolean formula θ and, depending on its value, selects either the second ρ_1 or the third argument ρ_2, where ρ_1 represents the loop body and ρ_2 represents either another ite operator, which encodes a k-copy of the loop body, or an assertion/assume statement.

Therefore, each step of our k-induction algorithm transforms a program with loops into a loop-free program, such that the correctness of the loop-free program implies the correctness of the program with loops.

If the program consists of multiple and possibly nested loops, we simply set the number of loop unwindings globally, that is, for all loops in the program and apply these aforementioned translations recursively. Note, however, that each case of the k-induction algorithm performs different transformations at the end of the loop: either to find bugs (base case) or to prove that enough loop unwindings have been done (forward condition).

7.3.1.3 Program Transformations

In terms of program transformations, which are all done completely automatically by our proposed method, the base case simply inserts an unwinding assumption, to the respective loop-free program P', consisting of the termination condition σ after the loop, as follows $I \wedge T \wedge \sigma \Rightarrow \phi$, where I is the initial condition, T is the transition relation of P', and ϕ is a safety property to be checked.

The forward case inserts an unwinding assertion instead of an assumption after the loop, as follows $I \wedge T \Rightarrow \sigma \wedge \phi$. The forward condition, proposed by Große et al. [18], introduces a sequence of commands to check whether there is a path between an initial state and the current state k, while in the algorithm proposed in this paper, an assertion (i.e., the loop invariant) is automatically inserted by our algorithm, without the user's intervention, at the end of the loop to check whether all states are reached in k steps. Our base case and forward condition translations can easily be implemented on top of plain BMC.

However, for the inductive step of the algorithm, several transformations are carried out. In particular, the loop $while(c)\,\{E;\,\}$ is converted into

$$A;\, while(c)\,\{S;\, E;\, U;\,\}\, R; \tag{7.2}$$

where A is the code responsible for assigning nondeterministic values to all loop variables, i.e., the state is havocked before the loop, c is the exit condition of the loop *while*, S is the code to store the current state of the program variables before executing the statements of E, E is the actual code inside the loop *while*, U is the code to update all program variables with local values after executing E, and R is the code to remove redundant states.

Definition 7.2 (*Loop Variable*) A loop variable is a variable $v \subseteq V$, where $V = V_{global} \cup V_{local}$ given that V_{global} is the set of global variables and V_{local} is the set of local variables that occur in the loop of a program.

Definition 7.3 (*Havoc Loop Variable*) A nondeterministic value is assigned to a loop variable v if and only if v is used in the loop termination condition σ, in the loop counter that controls iterations of a loop, or repeatedly modified inside the loop body.

The intuitive interpretation of S, U, and R is that if the current state (after executing E) is different than the previous state (before executing E), then new states are produced in the given loop iteration; otherwise, they are redundant and the code R is then responsible for preventing those redundant states to be included into the states vector. Note further that the code A assigns nondeterministic values to all loop variables so that the model checker can explore all possible states implicitly. In contrast, Große et al. [18] havoc all program variables, which makes it difficult to apply their approach to arbitrary programs since they do not provide enough information to constrain the havocked variables in the program. Similarly, the loop *for* can easily be converted into the loop *while* as follows: $for(B;\, c;\, D)\,\{E;\,\}$ is rewritten as

$$B;\ while(c)\,\{E;\ D;\,\} \tag{7.3}$$

where B is the initial condition of the loop, c is the exit condition of the loop, D is the increment of each iteration over B, and E is the actual code inside the loop *for*. No further transformations are applied to the loop *for* during the inductive step. Additionally, the loop *do while* can trivially be converted into the loop *while* with one difference, the code inside the loop must execute at least once before the exit condition is checked.

The inductive step is thus represented by $\gamma \wedge \sigma \Rightarrow \phi$, where γ is the transition relation of $\hat{P}'$, which represents a loop-free program (cf. Definition 7.1) after applying transformations (7.2) and (7.3). The intuitive interpretation of the inductive step is to prove that, for any unfolding of the program, there is no assignment of particular values to the program variables that violates the safety property being checked. Finally, the induction hypothesis of the inductive step consists of the conjunction between the postconditions ($Post$) and the termination condition (σ) of the loop.

7.3.1.4 Invariant Generation

To infer program invariants, we adopted the PIPS [24] tool, which is an interprocedural source-to-source compiler framework for C and Fortran programs and relies on a polyhedral abstraction of program behavior. PIPS has been developed for almost 20 years to analyze large size programs automatically [28]. PIPS performs a two-step analysis: (1) each program instruction is associated to an affine transformer, representing its underlying transfer function. This is a bottom-up procedure, starting from elementary instructions, then working on compound statements and up to function definitions; (2) polyhedral invariants are propagated along with instructions, using previously computed transformers.

In our proposed method, PIPS receives the analyzed program as input and then it generates invariants that are given as comments surrounding instructions in the output C code. These invariants are translated and instrumented into the program as assume statements. In particular, we adopt the function `assume` (*expr*) to limit possible values of the variables that are related to the invariants. This step is needed since PIPS generates invariants that are presented as mathematical expressions (e.g., $2j < 5t$), which are not accepted by C programs syntax and invariants with *#init* suffix that is used to distinguish the old value from the new value.

Algorithm 2 shows the method proposed, which receives as inputs the code generated by PIPS (`PIPSCode`) with invariants as comments, and it generates as output a new instance of the analyzed code (`NewCodeInv`) with invariants, adopting the function `assume` (*expr*), where *expr* is a expression supported by the C programming language. The time complexity of this algorithm is $O(n^2)$, where n is code size with invariants generated by PIPS. The algorithm is split into three parts: (1) identify the `#init` structure in the PIPS invariants; (2) generate code to support the translation of the `#init` structure in the invariant; and finally (3) translate mathematical expressions contained in the invariants, which is performed by the invariants transformation in the PIPS format to the C programming language.

Line 5 of Algorithm 2 performs the first part of the invariant translation, which consists of reading each line of the analyzed code with invariants and identifying whether a given comment is an invariant generated by PIPS (line 6). If an invariant is identified and it contains the structure `#init`, then the invariant location (the line number) is stored, as well as, the type and name of the variable, which has the structure prefix `#init` (line 8).

After identifying the `#init` structures in the invariants, the second part of Algorithm 2 performs line 12, which consists of reading again, each line of the analyzed code with invariants (`PIPSCode`), and identifying the beginning of each function in the code. For each identified function, the algorithm checks whether that function has identified some `#init` structure (line 15). If it has been identified, for each variable that has the suffix `#init`, a new line of code is generated at the beginning of the function, with the declaration of an auxiliary variable, which contains the old variable value, i.e., its value at the beginning of the function. The new created variable has the following format `type var_init = var`, where `type` is the identified variable type, and `var` is the identified variable name. During the execution of this algorithm, a new instance of the code (`NewCodeInv`) is generated.

In the third (and final part) of Algorithm 2 (line 22), each line of the new code instance (`NewCodeInv`) is read to convert each PIPS invariant into expressions supported by the C programming language. This transformation consists in applying regular expressions (line 27) to add operators (e.g., from $2j$ to $2 * j$) and replacing the structure `#init` to `_init`. For each analyzed PIPS comment/invariant, we generate a new line of code to the new format, where this line is concatenated with the operator `&&` and added to the `__ESBMC_assume` function.

Algorithm 2 Translation algorithm of PIPS invariants

```
Input: PIPSCode - C code with PIPS invariants
Output: NewCodeInv - New code with invariant supported by C programs
// dictionary to identify #init
dict_variniteloc ← { }
// list for the new code generated in the translation
NewCodeInv ← { }
// Part 1 - identifying #init in the invariants
for all line of the PIPSCode do
    if is a PIPS comment in this pattern // P(w,x) {w == 0, x#init > 10} then
        if the comment has the pattern ([a-zA-Z0-9_]+)#init then
            dict_variniteloc[line] ← the variable suffixed #init
        end if
    end if
end for
// Part 2 - code generation to support #init structure
for all line of PIPSCode do
    NewCodeInv ← line
    if is the beginning of a function then
        if has some line number of this function ∈ dict_variniteloc then
            for all variable ∈ dict_variniteloc do
                NewCodeInv ← Declare a variable with this pattern type var_init = var
            end for
        end if
    end if
end for
// Part 3 - correct the invariant format
for all line of NewCodeInv do
    // list to the translated invariants
    listinvpips ← { }
    NewCodeInv ← line
    if is a PIPS comment in this pattern // P(w,x) {w == 0, x#init > 10} then
        for all expression ∈ {w == 0, x#init > 10} do
            listinvpips ← Reformulate the expression according to the C programs syntax and
    replace #init by _init
        end for
        NewCodeInv ← __ESBMC_assume(concatenate the invariants in listinvpips with &&)
    end if
end for
```

7.3.2 Running Example

In this section, we explain how the k-induction algorithm (see Listing 25) can prove correctness of the C program shown in Listing 24.

7.3.2.1 The Base Case

The base case initializes the limits of the loop's termination condition with nondeterministic values so that the model checker can explore all possible states implicitly. The pre- and postconditions of the loop shown in Listing 24, in static single assignment (SSA) form [27], are as follows:

$$Pre := \begin{bmatrix} n_1 = nondet\_uint \wedge n_1 \geq 1 \\ \wedge\ sn_1 = 0 \wedge i_1 = 1 \end{bmatrix}$$

$$Post := \left[i_k \geq 1 \wedge\ i_k > n_1 \Rightarrow sn_k = n_1 \times a \right]$$

where Pre and $Post$ are the pre and postconditions to compute the sum given by Eq. (7.1), respectively, and $nondet\_uint$ is a nondeterministic function, which can return any value of type *unsigned int*. In the preconditions, n_1 represents the first assignment to the variable n, which is a nondeterministic value greater than or equal to one. This ensures that the model checker explores all possible unwindings of the program. Additionally, sn_1 represents the first assignment to the variable sn and i_1 is the initial condition of the loop. In the postconditions, sn_k represents the assignment $n + 1$ for the variable sn in Listing 24, which must be *true* if $i_k > n_1$. The code that is not pre or postcondition is represented by the variable Q (i.e., the sequence of commands inside the loop *for*) and it does not undergo any transformation during the base case. The resulting code of the base case transformations can be seen in Listing 26 (cf. Definition 7.1). Note that the *assume* (in line 11), which consists of the termination condition, eliminates all execution paths that do not satisfy the constraint $i > n$. This ensures that the base case finds a counterexample of depth k without reporting any false negative result. Note further that other assume statements, shown in Listing 24, are simply eliminated during the symbolic execution by propagating constants and checking that the resulting expression evaluates to *true* [12].

```
1  int main(int argc, char **argv) {
2    uint64_t i, sn=0;
3    uint32_t n=nondet_uint();
4      assume (n>=1);
5    i=1;
6    if (i<=n) {         \
7      sn = sn + a;       } k copies
8      i++;              /
9    }
10     ...
11   assume(i>n); // unwinding assumption
12   assert(sn==n*a);
13 }
```

Listing 26 Example code for the proof by mathematical induction - during base case.

7.3.2.2 The Forward Condition

In the forward condition, the k-induction algorithm attempts to prove that the loop is sufficiently unfolded and whether the property is valid in all states reachable within k steps. The postconditions of the loop shown in Listing 24, in SSA form, can then be defined as follows:

$$Post := \left[i_k > n_1 \wedge sn_k = n_1 \times a \right]$$

The preconditions of the forward condition are identical to the base case. In the postconditions $Post$, there is an assertion to check whether the loop is sufficiently expanded, represented by the constraint $i_k > n_1$, where i_k represents the value of the variable i at iteration $n + 1$. The resulting code of the forward condition transformations can be seen in Listing 27 (cf. Definition 7.1). The forward condition attempts to prove that the loop is unfolded deep enough (by checking the loop invariant in line 11) and if the property is valid in all states reachable within k iterations (by checking the assertion in line 12). As in the base case, we also eliminate assume expressions by checking whether they evaluate to *true* by propagating constants during symbolic execution.

```
1  int main(int argc, char **argv) {
2    uint64_t i, sn=0;
3    uint32_t n=nondet_uint();
4      assume (n>=1);
5    i=1;
6    if (i<=n) {          }
7      sn = sn + a;       } k copies
8      i++;               }
9    }
10     ...
11   assert(i>n); // check loop invariant
12   assert(sn==n*a);
13 }
```

Listing 27 Example code for the proof by mathematical induction - during forward condition.

7.3.2.3 The Inductive Step

In the inductive step, the k-induction algorithm attempts to prove that, if the property is valid up to depth k, the same must be valid for the next value of k. Several changes are performed in the original code during this step. First, a structure called *statet* is defined, containing all variables within the loop and the exit condition of that loop. Then, a variable of type *statet* called *cs* (current state) is declared, which is responsible for storing the values of a given variable in a given iteration; in the current implementation, the *cs* data structure does not handle heap-allocated objects. A state vector of size equal to the number of iterations of the loop is also declared, called sv (state vector) that will store the values of all variables on each iteration of the loop.

Before starting the loop, all loop variables (cf. Definitions 7.2 and 7.3) are initialized to nondeterministic values and stored in the state vector on the first iteration of the loop so that the model checker can explore all possible states implicitly. Within the loop, after storing the current state and executing the code inside the loop, all state variables are updated with the current values of the current iteration. An *assume* instruction is inserted with the condition that the current state is different from the previous one, to prevent redundant states to be inserted into the state vector; in this case, we compare $sv_j\,[i]$ to cs_j for $0 < j \leq k$ and $0 \leq i < k$. In the example we add constraints as follows:

$$\begin{aligned}
&sv_1\,[0] \neq cs_1\\
&sv_1\,[0] \neq cs_1 \wedge sv_2\,[1] \neq cs_2\\
&\ldots\\
&sv_1\,[0] \neq cs_1 \wedge sv_2\,[1] \neq cs_2 \wedge \ldots sv_k\,[i] \neq cs_k
\end{aligned} \tag{7.4}$$

Although we can compare $sv_k\,[i]$ to all cs_k for $i < k$ (since inequalities are not transitive), we found the encoding shown in Eq. (7.4) to be more efficient, leading to fewer timeouts when applied to the SV-COMP benchmarks.

Finally, an *assume* instruction is inserted after the loop, which is similar to that inserted in the base case. The pre- and postconditions of the loop shown in Listing 24, in SSA form, are defined as follows:

$$Pre := \begin{bmatrix} n_1 = nondet\_uint \wedge n_1 \geq 1 \\ \wedge\ sn_1 = 0 \wedge i_1 = 1 \\ \wedge\ cs_1.v_0 = nondet\_uint \\ \wedge\ \ldots \\ \wedge\ cs_1.v_m = nondet\_uint \end{bmatrix}$$

$$Post := \left[i_k > n_1 \Rightarrow sn_k = n_\times a \right]$$

In the preconditions Pre, in addition to the initialization of the variables, the value of all variables contained in the current state cs must be assigned with nondeterministic values, where m is the number of (automatic and static) variables that are used in the program. The postconditions do not change, as in the base case; they only contain the property that the algorithm is trying to prove. In the instruction set Q, changes are made in the code to save the value of the variables before and after the current iteration i, as follows:

$$Q := \begin{bmatrix} sv[i-1] = cs_i \wedge S \\ \wedge\ cs_i.v_0 = v_{0i} \\ \wedge\ \ldots \\ \wedge\ cs_i.v_m = v_{mi} \end{bmatrix}$$

In the instruction set Q, $sv[i-1]$ is the vector position to save the current state cs_i, S is the actual code inside the loop, and the assignments $cs_i.v_0 = v_{0i} \wedge \ldots \wedge cs_i.v_m = v_{mi}$ represent the value of the variables in iteration i being saved in the current state cs_i. The modified code for the inductive step, using the notation defined in Sect. 7.3.1, can be seen in Listing 28. Note that the *if*-statement (lines 18–26) in Listing 28 is copied k-times according to Definition 7.1. As in the base case, the inductive step also inserts an *assume* instruction, which contains the termination condition. Differently from the base case, the inductive step proves that the property, specified by the assertion, is valid for any value of n.

Lemma 7.1 *If the induction hypothesis* $\{Post \;\wedge\; \neg(i \leq n)\}$ *holds for* $k+1$ *consecutive iterations, then it also holds for k preceding iterations.*

After the loop *while* is finished, the induction hypothesis $\{Post \;\wedge\; \neg(i \leq n)\}$ is satisfied on any number of iterations; in particular, the SMT solver can easily verify Lemma 7.1 and conclude that $sn == n * a$ is inductive relative to n. As in previous cases, we also eliminate assume expressions by checking whether they evaluate to *true* by propagating constants during symbolic execution.

```
1  //variables inside the loop
2  typedef struct state {
3    long long int i, sn;
4    unsigned int n;
5  } statet;
6  int main(int argc, char **argv) {
7    uint64_t i, sn=0;
8    uint32_t n=nondet_uint();
9    assume (n>=1);
10   i=1;
11   //declaration of current state
12   //and state vector
13   statet cs, sv[n];
14   //A: assign nondeterministc values
15   cs.i=nondet_uint();
16   cs.sn=nondet_uint();
17   cs.n=n;
18   if (i<=n) { //c: exit condition
19     sv[i-1]=cs;  //S: store current state
20     sn = sn + a; //E: code inside the loop       } k copies
21     //U: update variables with local values
22     cs.i=i; cs.sn=sn; cs.n=n;
23     //R: remove redundant states
24     assume(sv[i-1]!=cs);
25     i++;
26   }
27   ...
28   assume(i>n); //unwinding assumption
29   assert(sn==n*a);
30 }
```

Listing 28 Example code for the proof by mathematical induction - during inductive step.

7.4 Experimental Evaluation

This section is split into two parts. The setup is described in Sects. 7.4.1, 7.4.2 describes a comparison among DepthK,[2] ESBMC [12], CBMC [23], and CPAchecker (Configurable Software Verification Platform) [7] using a set of C benchmarks from SV-COMP [4] and embedded applications [26, 30, 33].

7.4.1 Experimental Setup

The experimental evaluation is conducted on a computer with Intel Xeon CPU $E5-2670$ CPU, 2.60GHz, 115GB RAM with Linux $3.13.0-35$-generic x86_64. Each verification task is limited to a CPU time of 15 minutes and a memory consumption of 15 GB. Additionally, we defined the *max_iterations* to 100 (cf. Listing 25). To evaluate all tools, we initially adopted: 142 ANSI-C programs of the SV-COMP 2015 benchmarks[3]; in particular, the *Loops* subcategory; and 34 ANSI-C programs used in embedded systems: Powerstone [30] contains a set of C programs for embedded systems (e.g., for automobile control and fax applications); while SNU real time [33] contains a set of C programs for matrix and signal processing functions such as matrix multiplication and decomposition, quadratic equations solving, cyclic redundancy check, fast fourier transform, LMS adaptive signal enhancement, and JPEG encoding; and the WCET [26] contains C programs adopted for worst-case execution time analysis. Additionally, we present a comparison with the tools:

- DepthK v1.0 with k-induction and invariants using polyhedra, the parameters are defined in the wrapper script available in the tool repository;
- ESBMC v1.25.2 adopting k-induction without invariants. We adopted the wrapper script from SV-COMP 2013[4] to execute the tool;
- CBMC v5.0 with k-induction, running the script provided in [5];
- CPAChecker[5] with k-induction and invariants at revision 15596 from its SVN repository. The options to execute the tool are defined in [5]. To improve the presentation, we report only the results of the options that presented the best results. These options are defined in [5] as follows: CPAchecker *cont.-ref. k-Induction* (*k-Ind InvGen*) and CPAchecker *no-inv k-Induction*.

[2]https://github.com/hbgit/depthk.

[3]http://sv-comp.sosy-lab.org/2015/.

[4]http://sv-comp.sosy-lab.org/2013/.

[5]https://svn.sosy-lab.org/software/cpachecker/trunk.

7.4.2 Experimental Results

In preliminary tests with the DepthK, for programs from the SV-COMP 2015 loops subcategory, we observed that 4.92% of the results are false incorrect. We believe that, in turns, this is due to the inclusion of invariants, which over-approximates the analyzed program, resulting in incorrect exploration of the states sets. We further identify that, in order to improve the approach implemented in DepthK tool, ones needs to apply a rechecking/refinement of the result found by the BMC procedure. Here, we re-check the results using the forward condition and the inductive step of the k-induction algorithm.

In DepthK, the program verification with invariants modifies the k-induction algorithm (Listing 25), as presented in Algorithm 3. In this new k-induction algorithm, we added the following variables: `last_result`, which stores the last result identified in the verification of a given step of the k induction, and `force_basecase`, which is an identifier to apply the rechecking procedure in the base case of the k-induction. The main difference in the execution of Algorithm 3 is to identify whether in the forward condition (line 18) and the inductive step (line 22), the verification result was `TRUE`, i.e., there was no property violation in a new k unwindings.

After running all tools, we obtained the results shown in Table 7.1 for the SV-COMP 2015 benchmark and in Table 7.2 for the embedded systems benchmarks, where each row of these tables means: name of the tool (Tool); total number of programs that satisfy the specification (correctly) identified by the tool (Correct Results); total number of programs that the tool has identified an error for a program that meets the specification, i.e., false alarm or incomplete analysis (False Incorrect); total number of programs that the tool does not identify an error, i.e., bug missing or weak analysis (True Incorrect); Total number of programs that the tool is unable to model check due to lack of resources, tool failure (crash), or the tool exceeded the verification time of 15 min (Unknown and TO); the run time in minutes to verify all programs (Time).

Table 7.1 Experimental results for the SV-COMP'15 loops subcategory

Tool	DepthK	ESBMC + k-induction	CPAchecker no-inv k-Induction	CPAchecker cont.-ref. k-Induction (k-Ind InvGen)	CBMC + k-induction
Correct results	94	70	78	76	64
False incorrect	1	0	0	1	3
True incorrect	0	0	4	7	1
Unknown and TO	47	72	60	58	74
Time (min)	190.38	141.58	742.58	756.01	1141.17

Algorithm 3 The k-induction algorithm with a recheck in base case.

```
Input: Program P' with invariants and the safety proprities φ
Output: TRUE, FALSE, or UNKNOWN
k = 1
last_result = UNKNOWN
force_basecase = 0
while k <= max_iterations do
    if force_basecase > 0 then
        k=k+5
    end if
    if BASECASE(P', φ, k) then
        show the counterexample s[0 . . . k]
        return FALSE
    else
        if force_basecase > 0 then
            return last_result
        end if
        k=k+1
        if FORWARDCONDITION(P', φ, k) then
            force_basecase = 1
            last_result = TRUE
        else
            if INDUTIVESTEP(P', φ, k) then
                force_basecase = 1
                last_result = TRUE
            end if
        end if
    end if
end while
return UNKNOWN
```

Table 7.2 Experimental results for the Powerstone, SNU, and WCET benchmarks

Tools	DepthK	ESBMC + k-induction	CPAchecker no-inv k-Induction	CPAchecker cont.-ref. k-Induction (k-Ind InvGen)	CBMC + k-induction
Correct results	17	18	27	27	15
False incorrect	0	0	0	0	0
True incorrect	0	0	0	0	0
Unknown and TO	17	16	7	7	19
Time (min)	77.68	54.18	1.8	1.95	286.06

We evaluated the experimental results as follows: for each program we identified the verification result and time. We adopted the same scoring scheme that is used in SV-COMP 2015.[6] For every bug found, 1 score is assigned, for every correct

[6]http://sv-comp.sosy-lab.org/2015/rules.php.

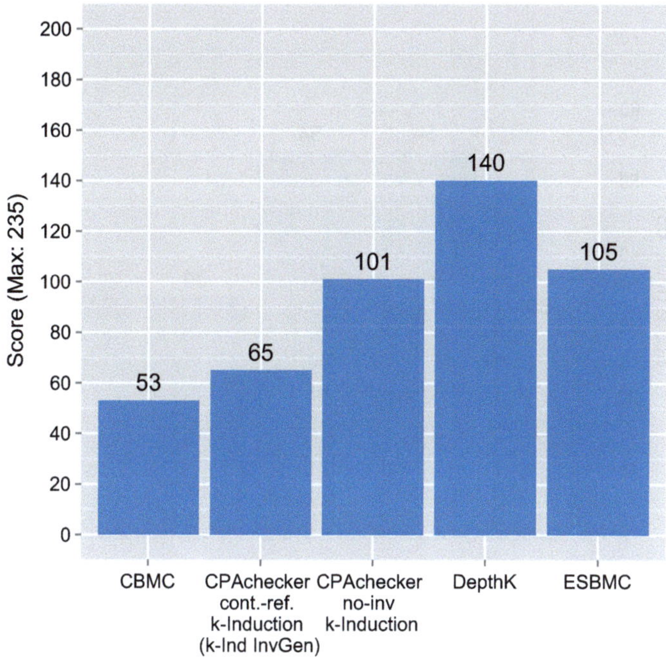

Fig. 7.1 Score to loops subcategory

safety proof, 2 scores are assigned. A score of 6 is subtracted for every wrong alarm (False Incorrect) reported by the tool, and 12 scores are subtracted for every wrong safety proof (True Incorrect). According to [6], this scoring scheme gives much more value in proving properties than finding counterexamples, and significantly punishes wrong answers to give credibility for tools. Figures 7.1 and 7.2 present the comparative results for the SV-COMP and embedded systems benchmarks, respectively. It is noteworthy that for the embedded systems programs, we have used safe programs [12] since we intend to check whether we have produced strong invariants to prove properties.

The experimental results in Fig. 7.1 show that the best scores belong to the DepthK, which combines k-induction with invariants, achieving 140 scores, ESBMC with k-induction without invariants achieved 105 scores, and CPAchecker *no-inv k-induction* achieved 101 scores. In Fig. 7.2, we found that the best scores belong to the CPAchecker *no-inv k-induction* with 54 scores, ESBMC with k-induction without invariants achieved 36 scores, and DepthK combined with k-induction and invariants, achieved 34 scores. We observed that DepthK achieved a lower score in the embedded system benchmarks. However, the DepthK results are still higher than that of CBMC and in the SV-COMP benchmark, DepthK achieved the highest score among all tools. In DepthK, we identified that, in turns, the lower score in the embedded system benchmarks is due to 35.30% of the results identified as `Unknown`, i.e., when

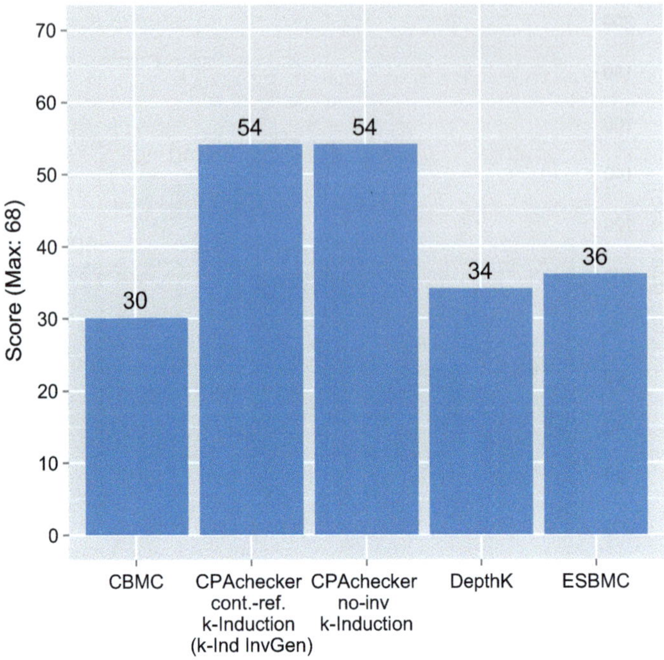

Fig. 7.2 Score to embedded systems

it is not possible to determine an outcome or due to a tool failure. We also identified failures related to invariant generation and code generation that is given as input to the BMC procedure. It is noteworthy that DepthK is still under development (in a somewhat preliminary state), so we argue that the results are promising.

To measure the impact of applying invariants to the k-induction based verification, we classified the distribution of the DepthK and ESBMC results, per verification step, i.e., base case, forward condition, and inductive step. Additionally, we included the verification tasks that result in `unknown` and `timeout` (CPU time exceeded 900 seconds). In this analysis, we evaluate only the results of DepthK and ESBMC, because they are part of our solution, and also because in the other tools, it is not possible to identify the steps of the k-induction in the standard logs generated by each tool. Figure 7.3 shows the distribution of the results, for each verification step, to the SV-COMP loops subcategory, and Fig. 7.4 presents the results to the embedded systems benchmarks.

The distribution of the results in Figs. 7.3 and 7.4 shows that DepthK can prove more than 25.35 and 29.41% of properties, during the inductive step, than ESBMC, respectively. These results lead to the conclusion that invariants helped the k-induction algorithm to prove that loops were sufficiently unwound and whenever the property is valid for k unwindings, it is also valid after the next unwinding of the system. We also identified that DepthK did not find a solution in 33.09% of the

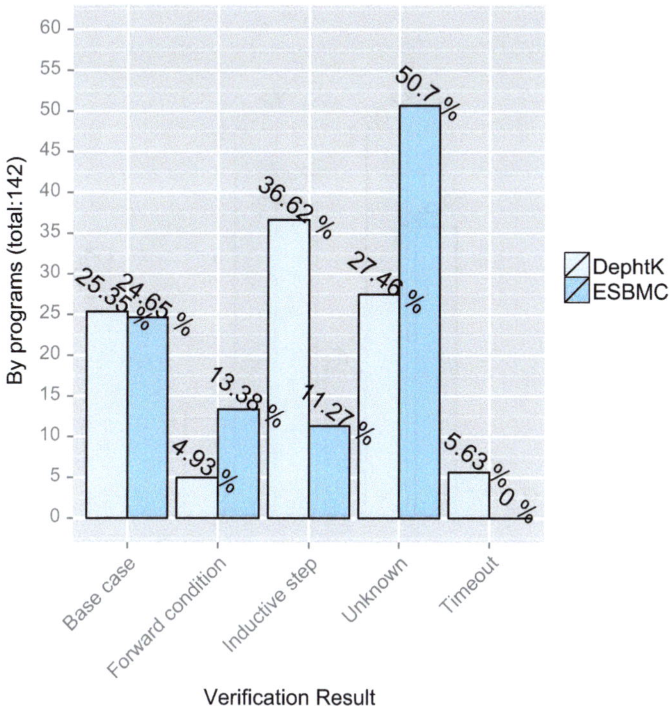

Fig. 7.3 Results for loops

programs in Fig. 7.3, and 50% in Fig. 7.4 (adding `Unknown` and `Timeout`). This is explained by the invariant generated from PIPS, which could not generate invariants strong enough to the verification with the k-induction, either due to a transformer or due to the invariants that are not convex; and also due to some errors in the tool implementation. ESBMC with k-induction did not find a solution in 50.7% of the programs in Fig. 7.3, i.e., 17.61% more than DepthK (adding `Unknown` and `Timeout`); and in Fig. 7.4, ESBMC did not find a solution in 47.06%, then only 3.64% less than the DepthK, thus providing evidences that the program invariants combined with k-induction can improve the verification results.

In Table 7.1, the verification time of DepthK to the loops subcategory is typically faster than the other tools, except for ESBMC, as can be seen in Fig. 7.5. This happens because DepthK has an additional time for the invariants generation. In Table 7.2, we identified that the verification time of DepthK is only faster than CBMC, as shown in Fig. 7.6. However, note that the DepthK verification time is proportional to ESBMC, since the time difference is 23.5 min; we can argue that this time difference is associated to the DepthK invariant generation.

We believe that the DepthK verification time can be significantly improved in two directions: fix some errors in the tool implementation, because some results generated as `Unknown` are related to failures in the tool execution; and adjustments in the PIPS

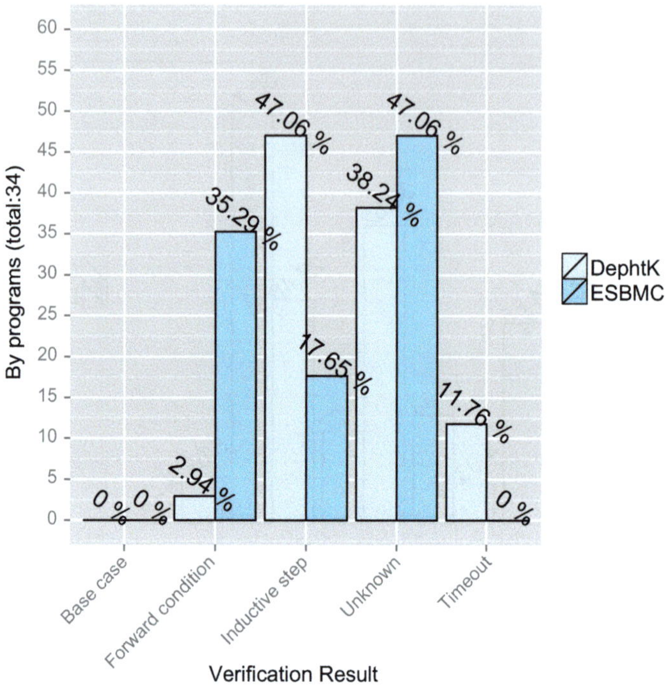

Fig. 7.4 Results for embedded programs

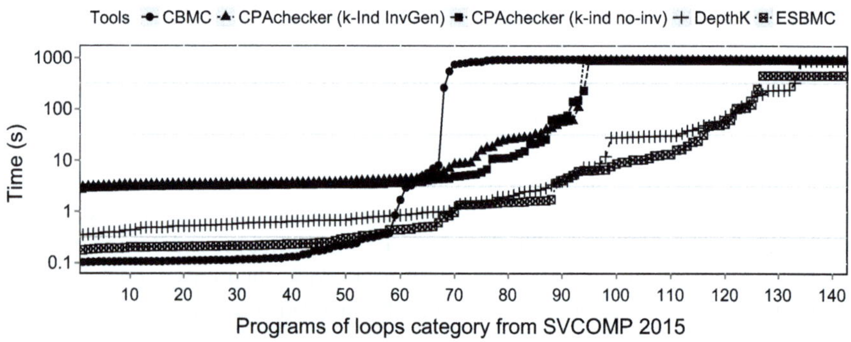

Fig. 7.5 Verification time to the loops subcategory

script parameters to generate invariants, since PIPS has a broad set of commands for code transformation, which might have a positive impact in the invariant generation for specific class of programs.

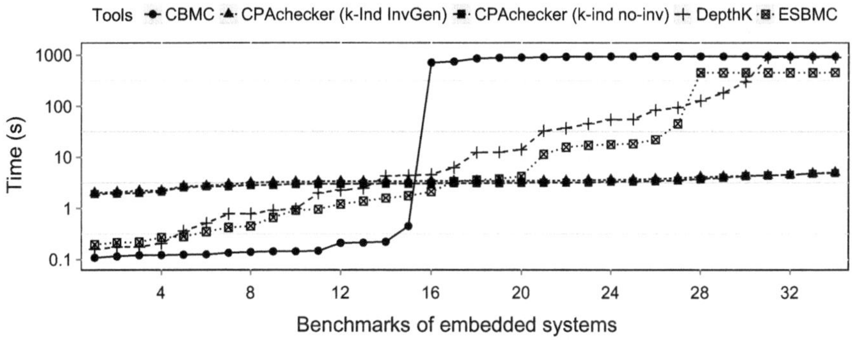

Fig. 7.6 Verification time to the embedded systems programs

7.5 Related Work

The application of the k-induction method is gaining popularity in the software verification community. Recently, Bradley et al. introduce "property-based reachability" (or IC3) procedure for the safety verification of systems [9, 20]. The authors have shown that IC3 can scale on certain benchmarks where k-induction fails to succeed. However, we do not compare k-induction against IC3 since it is already done by Bradley [9]; we focus our comparison on related k-induction procedures.

Previous work on the one hand have explored proofs by mathematical induction of hardware and software systems with some limitations, e.g., requiring changes in the code to introduce loop invariants [14, 15, 18]. This complicates the automation of the verification process, unless other methods are used in combination to automatically compute the loop invariant [1, 31]. Similar to the approach proposed by Hagen and Tinelli [19], our method is completely automatic and does not require the user to provide loops invariants as the final assertions after each loop. On the other hand, state-of-the-art BMC tools have been widely used, but as bug-finding tools since they typically analyze bounded program runs [11, 25]; completeness can only be ensured if the BMC tools know an upper bound on the depth of the state space, which is not generally the case. This paper closes this gap, providing clear evidence that the k-induction algorithm can be applied to a broader range of C programs without manual intervention.

Große et al. describe a method to prove properties of TLM designs (Transaction Level Modeling) in SystemC [18]. The approach consists of converting a SystemC program into a C program, and then it performs the proof of the properties by mathematical induction using the CBMC tool [11]. The difference to the one described in this paper lies on the transformations carried out in the forward condition. During the forward condition, transformations similar to those inserted during the inductive step in our approach, are introduced in the code to check whether there is a path between an initial state and the current state k; while the algorithm proposed in this paper, an assertion is inserted at the end of the loop to verify that all states are reached in k steps.

Donaldson et al. describe a verification tool called Scratch [15] to detect data races during Direct Memory Access (DMA) in the CELL BE processor from IBM [15]. The approach used to verify C programs is the k-induction technique. The approach was implemented in the Scratch tool that uses two steps, the base case and the inductive step. The tool is able to prove the absence of data races, but it is restricted to verify that specific class of problems for a particular type of hardware. The steps of the algorithm are similar to the one proposed in this paper, but it requires annotations in the code to introduce loops invariants.

Kahsai et al. describe PKIND, a parallel version of the tool KIND, used to verify invariant properties of programs written in Lustre [22]. In order to verify a Lustre program, PKIND starts three processes, one for base case, one for inductive step, and one for invariant generation, note that unlike ESBMC, the k-induction algorithm used by PKIND does not have a forward condition step. This because of PKIND is for Lustre programs that do not terminate. Hence, there is no need for checking whether loops have been unrolled completely. The base case starts the verification with $k = 0$, and increments its value until it finds a counterexample or it receives a message from the inductive step process that a solution was found. Similarly, the inductive step increases the value of k until it receives a message from the base case process or a solution is found. The invariant generation process generates a set of candidates invariants from predefined templates and constantly feeds the inductive step process, as done recently by Beyer et al. [6].

7.6 Conclusions

The main contributions of this work are the design, implementation, and evaluation of the k-induction algorithm, adopting invariants using polyhedra in a verification tool, as well as, the use of the technique for the automated verification of reachability properties in embedded systems programs. To the best of our knowledge, this paper marks the first application of the k-induction algorithm to a broader range of embedded C programs. To validate the k-induction algorithm, experiments were performed involving 142 benchmarks of the SV-COMP 2015 *loops* subcategory, and 34 ANSI-C programs from the embedded systems benchmarks. Additionally, we presented a comparison to the ESBMC with k-induction, CBMC with k-induction, and CPAChecker with k-induction and invariants.

The experimental results are promising; the proposed method adopting k-induction with invariants (implemented in DepthK tool) determined 11.27% more accurate results than that obtained by CPAChecker, which had the second best result in the SV-COMP 2015 loops subcategory. The experimental results also show that the k-induction algorithm without invariants was able to verify 49.29% of the programs in the SV-COMP benchmarks in 141.58 min, and k-induction with invariants using polyhedra (i.e., DepthK) was able to verify 66.19% of the benchmarks in 190.38

min. Therefore, we identified that k-induction with invariants determined 17% more accurate results than the k-induction algorithm without invariants.

For embedded systems benchmarks, we identified some improvements in the DepthK tool, related to defects in the tool execution, and possible adjustments to invariant generation with PIPS. This is because the results were inferior compared to the other tools for the embedded systems benchmarks, where DepthK only obtained better results than CBMC tool. However, we argued that the proposed method, in comparison to other state-of-the-art tools, showed promising results indicating its effectiveness. In addition, both forms of the proposed method were able to prove or falsify a wide variety of safety properties; however, the k-induction algorithm, adopting polyhedral solves more verification tasks, which demonstrate an improvement of the induction k-algorithm effectiveness.

References

1. Ancourt C, Coelho F, Irigoin F (2010) A modular static analysis approach to affine loop invariants detection. In: Electronic notes in theoretical computer science (ENTCS). Elsevier Science Publishers B. V, pp 3–16
2. Barrett CW, Sebastiani R, Seshia SA, Tinelli C (2009) Satisfiability modulo theories. In: Handbook of satisfiability. IOS Press, pp 825–885
3. Beyer D (2013) Second competition on software verification—(Summary of SV-COMP 2013). In: Conference on tools and algorithms for the construction and analysis of systems (TACAS). Springer, pp 594–609
4. Beyer D (2015) Software verification and verifiable witnesses—(Report on SV-COMP 2015). In: Conference on tools and algorithms for the construction and analysis of systems (TACAS), pp 401–416
5. Beyer D, Dangl M, Wendler P (2015) Boosting k-Induction with continuously-refined invariants. http://www.sosy-lab.org/~dbeyer/cpa-k-induction/
6. Beyer D, Dangl M, Wendler P (2015) Combining k-Induction with continuously-refined invariants. CoRR abs/1502.00096. http://arxiv.org/abs/1502.00096
7. Beyer D, Keremoglu ME (2011) CPAchecker: a tool for configurable software verification. In: Conference on computer-aided verification (CAV), pp 184–190
8. Biere A (2009) Bounded model checking. In: Handbook of satisfiability. IOS Press, pp 457–481
9. Bradley AR (2012) IC3 and beyond: incremental, inductive verification. In: Computer aided verification (CAV). Springer, p 4
10. Brummayer R, Biere A (2009) Boolector: an efficient SMT solver for bit-vectors and arrays. In: Proceedings of the 15th international conference on tools and algorithms for the construction and analysis of systems: held as part of the joint European conferences on theory and practice of software (ETAPS). Springer, pp 174–177
11. Clarke E, Kroening D, Lerda F (2004) A tool for checking ANSI-C programs. In: Tools and algorithms for the construction and analysis of systems (TACAS). Springer, pp 168–176
12. Cordeiro L, Fischer B, Marques-Silva J (2012) SMT-based bounded model checking for embedded ANSI-C software. IEEE Trans Softw Eng (TSE):957–974
13. De Moura L, Bjørner N (2008) Z3: an efficient SMT solver. In: Tools and algorithms for the construction and analysis of systems (TACAS). Springer, pp 337–340
14. Donaldson AF, Haller L, Kroening D, Rümmer P (2011) Software verification using k-induction. In: Proceedings of the 18th international static analysis symposium (SAS). Springer, pp 351–368

15. Donaldson AF, Kroening D, Ruemmer P (2010) Automatic analysis of scratch-pad memory code for heterogeneous multicore processors. In: Proceedings of the 16th international conference on tools and algorithms for the construction and analysis of systems (TACAS). Springer, pp 280–295
16. Eén N, Sörensson N (2003) Temporal induction by incremental SAT solving. Electronic notes in theoretical computer science (ENTCS), pp 543–560
17. Gadelha M, Ismail H, Cordeiro L (2015) andling loops in bounded model checking of c programs via k-induction. Int J Softw Tools Technol Transf (to appear) (2015)
18. Große D, Le HM, Drechsler R (2009) Induction-based formal verification of SystemC TLM designs. In: 10th International workshop on microprocessor test and verification (MTV), pp 101–106
19. Hagen G, Tinelli C (2008) Scaling up the formal verification of Lustre programs with SMT-based techniques. In: Proceedings of the 8th international conference on formal methods in computer-aided design (FMCAD). IEEE, pp 109–117
20. Hassan Z, Bradley AR, Somenzi F (2013) Better generalization in IC3. In: Formal methods in computer-aided design (FMCAD). IEEE, pp 157–164
21. Ivancic F, Shlyakhter I, Gupta A, Ganai MK (2005) Model checking C programs using F-SOFT. In: 23rd international conference on computer design (ICCD). IEEE Computer Society, pp 297–308
22. Kahsai T, Tinelli C (2011) Pkind: A parallel k-induction based model checker. In: Proceedings 10th international workshop on parallel and distributed methods in verification (PDMC), pp 55–62
23. Kroening D, Tautschnig M (2014) CBMC—C Bounded model checker—(Competition Contribution). In: Conference on tools and algorithms for the construction and analysis of systems (TACAS), pp 389–391
24. Maisonneuve V, Hermant O, Irigoin F (2014) Computing invariants with transformers: experimental scalability and accuracy. In: 5th International workshop on numerical and symbolic abstract domains (NSAD). Electronic notes in theoretical computer science (ENTCS). Elsevier, pp 17–31
25. Merz F, Falke S, Sinz C (2012) LLBMC: bounded model checking of C and C++ programs using a compiler IR. In: Proceedings of the 4th international conference on verified software: theories, tools, experiments (VSTTE). Springer, pp 146–161
26. MRTC: WCET Benchmarks (2012) Mälardalen Real-Time Research Center. http://www.mrtc.mdh.se/projects/wcet/benchmarks.html
27. Muchnick SS (1997) Advanced compiler design and implementation. Morgan Kaufmann Publishers Inc., San Francisco, CA, USA
28. ParisTech M (2013) PIPS: Automatic parallelizer and code transformation framework. http://pips4u.org
29. Rocha H, Ismail H, Cordeiro LC, Barreto RS (2015) Model checking embedded C software using k-induction and invariants. In: Brazilian symposium on computing systems engineering (SBESC). IEEE, pp 90–95
30. Scott J, Lee LH, Arends J, Moyer B (1998) Designing the Low-Power M*CORE Architecture. In: Power driven microarchitecture workshop, pp 145–150
31. Sharma R, Dillig I, Dillig T, Aiken A (2011) Simplifying loop invariant generation using splitter predicates. In: Proceedings of the 23rd international conference on computer aided verification (CAV). Springer, pp 703–719
32. Sheeran M, Singh S, Stålmarck G (2000) Checking safety properties using induction and a SAT-solver. In: Formal methods in computer-aided design (FMCAD), pp 108–125
33. SNU (2012) SNU real-time benchmarks. http://www.cprover.org/goto-cc/examples/snu.html

Chapter 8
Scalable and Optimized Hybrid Verification of Embedded Software

Jörg Behrend, Djones Lettnin, Alexander Grünhage, Jürgen Ruf, Thomas Kropf and Wolfgang Rosenstiel

8.1 Introduction

Embedded software (ESW) is omnipresent in our daily life. It plays a key role in overcoming the time-to-market pressure and providing new functionalities. Therefore, a high number of users are dependent on its functionality [1]. ESW is often used in safety critical applications (e.g., automotive, medical, avionic), where correctness is of fundamental importance. Thus, verification and validation approaches are an important part of the development process.

The most commonly used approaches to verify embedded software are based on simulation or formal verification (FV). Testing, co-debugging and/or co-simulation techniques result in a tremendous effort to create test vectors. Furthermore, critical corner case scenarios might remain unnoticed. An extension of simulation is the assertion-based verification (ABV) methodology that captures a design's intended

J. Behrend (✉) · A. Grünhage · J. Ruf · T. Kropf · W. Rosenstiel
Department of Computer Engineering, University of Tübingen,
Sand 13, 72076 Tübingen, Germany
e-mail: behrend@informatik.uni-tuebingen.de

A. Grünhage
e-mail: gruenhag@informatik.uni-tuebingen.de

J. Ruf
e-mail: ruf@informatik.uni-tuebingen.de

T. Kropf
e-mail: kropf@informatik.uni-tuebingen.de

W. Rosenstiel
e-mail: rosenstiel@informatik.uni-tuebingen.de

D. Lettnin
Department of Electrical and Electronic Engineering, Federal University of Santa Catarina,
Campus Universitário s/n, Trindade, Florianópolis, SC CEP 88040-900, Brazil
e-mail: djones.lettnin@ufsc.br

D. Lettnin and M. Winterholer (eds.), *Embedded Software Verification and Debugging*, Embedded Systems, DOI 10.1007/978-1-4614-2266-2_8

behavior in temporal properties. This methodology has been successfully used at lower levels of hardware designs, which are not suitable for software. ESW has no timing reference and contains more complex data structures (e.g., integers, pointers) requiring a new mechanism to apply an assertion-based methodology. In order to verify temporal properties in ESW, formal verification techniques are efficient, but only up to medium sized software systems. For more complex designs, formal verification using model checking often suffers from the state space explosion problem. Therefore, abstraction techniques (e.g., predicate abstraction [2]) are applied to reduce the load of the back-end model checker.

Semiformal or hybrid approaches have been proposed many times before with only limited success. In this paper we present VERIFYR [3], an optimized and scalable hybrid verification approach using a semiformal algorithm and taking advantage of automated static parameter assignment (SPA). This technique reduces the model size by assigning a static value to at least one function parameter. Information gained during simulation (dynamic verification) is used to assign values to the parameters in order to reduce the formal model (static verification). One issue is the selection of the best function parameter. This is important due to the different impact of parameters on the resulting state space. Until now, it was a manual or randomized task to assign the parameter values. The selection of the parameter values may influence the program flow and therefore, the resulting model size. This work describes a new approach in order to rank function parameters depending on their impact on the model size. The ranking is based on estimation according to the usage of the parameters in the function body. Finally, SPA can be automatically applied to select parameters in an optimized way in order to reduce the model complexity in a controlled manner.

The paper is organized as follows. Section 8.2 describes the related work. Section 8.3 details the verification methodology and the technical details. Section 8.4 summarizes our case studies and presents the results. Section 8.5 concludes this paper and describes the future work.

8.2 Related Work

Bounded model checking (BMC) is an approach to reduce the model size using bounded execution paths. The key idea is to build a propositional formula, whose models correspond to program traces (with bounded length) that might violate some given property using state-of-the-art SAT and SMT solvers [4]. For instance, C bounded model checker (CBMC) [5–7] has proven to be a successful approach for automatic software analysis. Codeiro et al. [8] have implemented ESBMC based on the front-end of CBMC and a new back-end based on SMT. All the above-mentioned work fail when the bound is not automatically determinable.

The optimization of formal models has been the reason to use abstraction methods. The automatic predicate abstraction [9] introduced a way to construct abstracted models and allowed to introduce automated constraints like loop invariants [10]. Based on abstraction, Clarke et al. developed a refinement technique to generate even smaller models using counterexamples [11, 12]. BLAST [13] and SATABS [14] are formal verification tools for ANSI-C programs. Both tools use predicate abstraction mechanisms to enhance the verification process and to reduce the model size successfully. Semiformal/hybrid verification approaches have been applied successfully to hardware verification [15–17]. However, the application of a current semiformal hardware model checker to verify embedded software is not viable for large industrial programs [18]. In the area of embedded software using C language, Lettnin et al. [19] proposed a semiformal verification approach based on simulation and symbolic model checker (SymC) [20]. However, SymC was the bottleneck for the scalability of the formal verification, since it was originally developed for the verification of hardware designs. Cordeiro et al. [21] have published a semiformal approach to verify medical software, but they have scalability problems caused by the used model checker. The aforementioned related works have their pros and cons. However, they still have scalability limitations in the verification of complex embedded software with or without hardware dependencies.

Concolic testing was first introduced by Godefroid et al. [22] and Cadar et al. [23] independently. Koushik et al. [24] extended this methodology to a hybrid software verification technique mixing symbolic and concrete execution. They treat program variables as symbolic variables along a concrete execution path. Symbolic execution is used to generate new test cases to maximize the code coverage. The main focus is finding bugs, rather than proving program correctness. The resulting tools DART, EXE, and CUTE apply concolic testing to C programs. But concolic testing has problems when very large symbolic representations have to be generated, often resulting in unsolvable problems. Other problems like imprecise symbolic representations or incomplete theorem proving often result in a poor coverage. ULISSE [25] is a tool to support system-level specification testing based on extended finite state machines (EFSM). The KLEE [26] framework compiles the source code to LLVM [27] byte code. The code under test has to be compatible with LLVM and user interaction (which is essential for our verification approach) is not supported. PEX [28] was developed at Microsoft Research to automate structural testing of .NET code but not C code. Frama-C [29–31] is an integrated development environment for C code with focus on static verification only and Frama-C needs special code annotations for the used "design by contract" approach.

Behrend et al. used SPA [3] to reduce the model size during semiformal verification. By assigning a static value to a function parameter the automatic predicate abstraction algorithm can generate a different abstraction that may lead to a smaller model. If a parameter is assigned, the parameter is no longer handled as full range variable, but as statically assigned variable. However, the previous approach the function parameter for SPA was selected manually.

8.2.1 Contributions

Our main contribution in this current work is a novel semiformal approach for the verification of embedded software with temporal properties based on VERIFYR. We provide a new methodology to extract both dynamic and static verifiable models from C programs to perform both assertion-based and formal verification. On the formal side, we are able to extend the formal engine with different state-of-the-art software model checkers (SMC). On the simulation side, simulation models (C or SystemC) and the testbench environment can be automatically generated including randomization policies for input variables. Concerning our hybrid approach, on one hand, the formal verification is able to guide the simulation process based on the counterexamples. On the other hand, the simulation engine supports the formal verification, for instance, with the assignment of automated static parameters in order to shrink the state space. In previous work [3], the SPA was determined by hand or using a random selection, that is, a try-and-error method. In this work, we enable for the first time the automated assignment of static parameters via a new ranking algorithm with the following specific contributions:

- Automated SPA: An automatic usage of SPA is possible using this heuristic.
- Testbench: Automatic generation of testbenchs and simulation/formal models.
- Code quality and safety: The ranking can be used to detect dead parameters as well as high complex functions based on high rated parameters.
- Minimize time/costs: The effort of brute-forcing all parameters using SPA can be reduced by testing specific and promising parameters.
- Maximized coverage: Using this heuristic the smallest restriction and therefore the widest coverage can be determined.

8.3 VERIFYR Verification Methodology

Figure 8.1 and Algorithm 4 delineates the semiformal verification algorithm. Our approach is based on the analysis of the embedded software structure via a function call graph (FCG), as shown in Fig. 8.1a. The FCG represents the calling relationships between functions of embedded software. The verification strategy is divided in three phases: preprocessing (Algorithm 4, lines 2–8), formal exploration phase (a.k.a. bottom-up) (Algorithm 4, lines 10–17), and semiformal verification phase (a.k.a. top-down) (Algorithm 4, lines 18–29). In summary, the Formal Exploration (bottom-up) phase identifies which functions are too complex to be verified by standalone software model checkers. After identifying these functions we start the Semiformal (top-down) phase combining simulation, SPA and formal verification in order to overcome the software complexity.

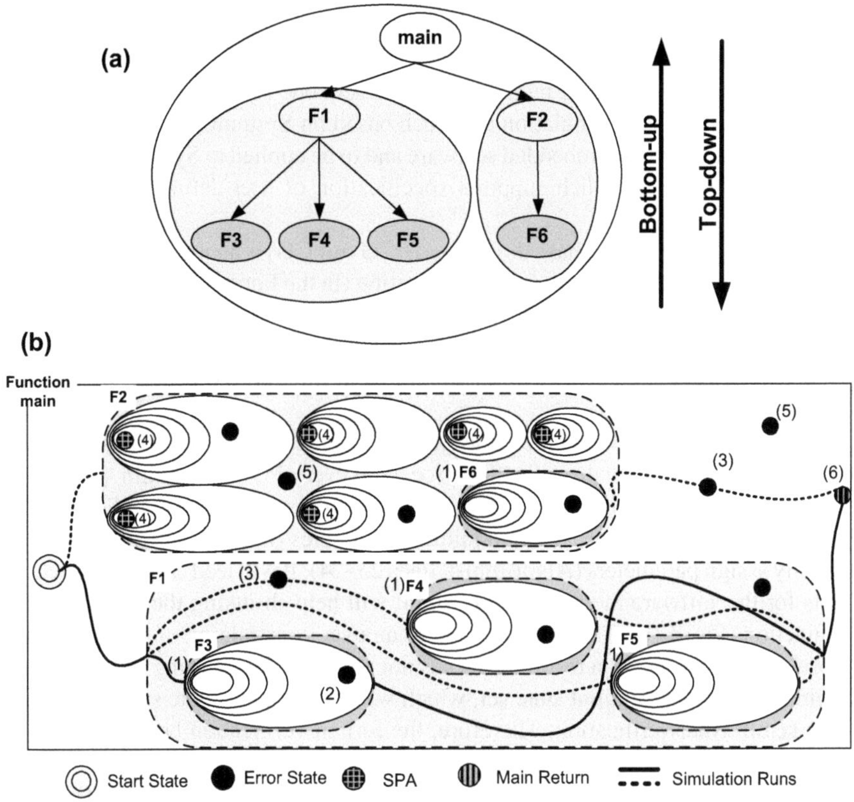

Fig. 8.1 VERIFYR verification approach

The VERIFYR verification methodology starts with the Formal Exploration phase. It uses state-of-the-art software model checkers (SMC) with built-in and user-defined properties specified in LTL to verify all functions of the FCG. It begins with the leaves (e.g., functions F3, F4, F5, F6 in Fig. 8.1) and continues in the next upper levels until the verification process reaches the function `main` (Algorithm 4, lines 11–12). If it is not possible to verify a function with the SMCs, it is marked in the FCG (Algorithm 4, line 13) (e.g., function `main`, F1, and F2 in Fig. 8.1). This means that these functions are too complex to be verified by a standalone model checker due to time out (TO) or out of memory (MO) constraints and that it is required to perform the semiformal/hybrid phase. Finally, a marked FCG (mFCG) is returned including all functions that failed during the Formal Exploration phase, however, if mFCG is empty then all function were formally verified and the verification process is completed (Algorithm 4, lines 15–16).

The Semiformal phase starts the simulation run with the assertion-based approach (ABV) (Algorithm 4, line 19), which requires one simulation model (or original C program wrapped in SystemC model) and testbench, one or many properties in LTL to be checked. We use a simulation approach based on SystemC. Thus, we derive a SystemC model from the embedded software and to be applied to SystemC Temporal Checker (SCTC) [32], which supports specification of user-defined properties in LTL [33].

During the Semiformal phase the marked FCG (mFCG) is analyzed. All functions that were not yet verified due to failed verification (in the Formal Exploration phase) are marked as point of interest (POI). POIs are basically the initial states of the local functions (F2 in Fig. 8.1b4). Therefore, the mFCG is used as a guiding mechanism in order to determine which function should be verified at the formal verification phase.

The Simulation engine monitors the simulation process (i.e., properties and variables) to start a new formal verification process at every POI (Algorithm 4, line 21). ABV is responsible for finding the POIs as well as the error states (F2 in Fig. 8.1b3). We use the monitored information to initialize variables (interaction with formal) to statically assign parameters (Algorithm 4, lines 23–24). It will lead to different access points for the software model checkers and it will help shrinking the state space of the function (Fig. 8.1b). This heuristic avoids an over-constraining of the state space in formal verification. As a result, SMC has not only a unique starting state (as usual by simulation), but an initial state set, which will improve the state space coverage of the semiformal verification. Therefore, the formal verification benefits from the simulation, as show in Fig. 8.1(F2).

Finally, a temporary version of the source code of the function under test (FUT) is created and is checked with the formal SMCs (Algorithm 4, lines 25). If a counterexample is reported, this information is used to guide the simulation (learning process). For instance, the randomization of input variables in our testbench is constrained in order to generate more efficient test vectors. Additionally, if desired, the user can set randomization constraints manually. Currently, when a counterexample should be reported to the user we save the global variable assignment of the used simulation run ("seed") to trace back from the counterexample given by the SMC to the entry point of the simulation run. Then we translate the CIL generated information back to the original C code.

When the simulation run reaches the `return` operation of the `main` function, a new simulation run is started. The global interaction between simulation and formal verification will continue until all the properties were evaluated or, a time bound or maximum number of simulation runs is reached or no more marked functions are available (Algorithm 4, line 20).

In the next sections, the SPA heuristic as well as the modeling details of embedded software will be presented.

Algorithm 4 VERIFYR algorithm

```
VERIFYR(Cprog,PropSet)
   doPreProcessing()
        C3AC = 3ACGen(Cprog)
        CTestbench = testbenchGen(Cprog)
        C3AC = propToAssertionSMC(C3AC, PropSet)
        CTestbench = propToAssertionSim(CTestbench,
  PropSet)
        FCG = FCGgen(C3AC)
   end doPreProcessing
   startOrchestrator()
        doFormalExploration() //BOTTOM-UP
              for each CFunction in C3AC
                   VStatus = startSMC(CFunction)
                   if VStatus == FAIL then
                        mFCG += markFunction(CFunction)
              if mFCG == NULL then
                   return VCOMPLETE
         end doFormalExploration
         doSemiformal() //TOP-DOWN
              startSimulation(CTestbench, Cprog)
            while NoTimeBound or NoMaxSimRuns or !Empty(
  mFCG)
                   POIFunction = watchSimFunctions(mFCG)
                   if POIFunction in mFCG then
           assessParameterScore(POIFunction)
           CFunction = doSPA(POIFunction)
     VStatus = startSMC(CFunction)
     if VStatus == COUNTEREXE  then
                  guideSimulation()
           unmarkFunction(mFCG)
   end doHybridFormalSimulation
   doComputeCoverage()
   doShowCounterexample()
   end startOrchestrator
end VERIFYR
```

8.3.1 SPA Heuristic

The SPA heuristic assumes that there is a function list containing all functions with all their parameters in a structured way. The algorithm iterates through the statements of each function body in the function list, inspecting each statement. If a statement contains one of the function parameters, this statement is inspected in more details. The analysis covers 11 aspects of the statement called properties. More details on these properties are in Sect. 8.3.1.2. Based on these properties the statement is assessed and a score is computed. The scores are summed up and stored for each parameter. The statement is examined for introducing a parameter value-dependent variable

(PVDV). This is done after the property check since the first statement of this PVDV is not rated. They are queued in the parameter list marked with their depending parameter. The achieved score of a PVDV is added to the score of the parameter the PVDV depends on.

8.3.1.1 Parameter Value-Dependent Variables

Variables that are initialized using a parameter are directly affected by SPA. This observation led to the concept of PVDV. Applying SPA on a function parameter reduces the model size because the model is not required to cover the full range of possible values. This effect is passed down to PVDV as they are directly depending on the parameter value. They passed the effect on to their value-dependent variables.

In order to cover this effect this technique monitors the value dependencies by analyzing assignments. If a PVDV is found, the variable is queued in the list of parameters. Any impact on the model size (such as PVDV) is added to the impact of the function parameter on which value the PVDV was initialized. Keeping track of these PVDVs is an essential part of this technique. This is because it is a common practice to make copies of parameters if those are used at multiple locations. And the parameters that are used in multiple locations have a huge impact.

8.3.1.2 Context Properties

Every statement that contains a parameter is evaluated against a set of internal properties. These properties describe the context in which the parameter is used within the statement. Therefore, the properties cover all context aspects of a statement that are used to state an assessment. All properties are determined by inspecting the code and are the base for the later assessment. Following eleven properties reflect the aspects of a statement regarding the usage as a variable:

- **Reading**: True if the parameter is on the right hand side of an assignment.
- **Writing**: True if the parameter is on the left hand side of an assignment.
- **Compare**: True if there is a comparison.
- **Loop**: True if statement contains the keyword “for” or “while.”
- **Function**: True if the parameter is in brackets, as it would be when used as function parameter.
- **Conditional**: True if the statement contains the keyword “if,” “switch,” or “case.”
- **Return**: True if the statement starts with the keyword “return.”
- **Command**: True if the statement ends with a semicolon.
- **Multiple use**: True if the statement contains one parameter multiple times.
- **Indirect use**: True if the monitored parameter is not a direct parameter but a PVDV.
- **First use**: True only at the first appearance of a parameter.

8.3.1.3 Assessing Function

The assess function estimates the model size based on the usage of a parameter in a statement. The estimation is based on the properties and is implemented as a Boolean clause. The number of cases with significant impact on the model size is limited. In this heuristic the four following special cases are used:

- **Dead parameter**: If a parameter is written on the first appearance the parameter is considered dead. SPA may already be applied.
- **Return**: Actual function parameters (not PVDV) that are returned have lesser impact.
- **Switch statement**: Conditional parameters control the program flow and therefore, impact heavily on the model size.
- **Loop boundary**: Loops are commonly unwound within the formal model, so the loop boundary has a major impact on the model size.

Each case is rewarded with a score. The number of points per case is reflecting the impact on the model size. As every statement has a basic impact every statement receives one point. If variations of a parameter value do not impact the model size, the parameter is called a dead parameter. These dead parameters are mainly parameters that already have SPA applied. This case is rewarded with a negative score to lower the ranking to a minimum. In addition dead parameters are not assessed anymore. Return parameters are mainly data containers that have lesser impact on the model size than not returned parameters. This is an observation made while testing the heuristic on the available benchmark. An implementation should use a low negative score to cover this effect. Parameters that are used in conditional statements have great impact on the model size as they control the program flow. The score should be set to a high value to ensure that parameters that are not used in conditional or loop statements cannot reach a higher rank. Loop boundary parameters are parameters that control the boundary of a loop. As loops are commonly unwound in the formal model the impact of those parameters on the model size are huge. Experiments using the available benchmark showed that the impact on the model size of three conditional statements can surpass the impact of one loop boundary parameter. In order to cover this fact the score should be set to twice the score of a conditional statement.

8.3.2 Preprocessing Phase

Basically the preprocessing phase considers the generation of testbench and simulation models, preprocessing the C program to the software model checker, defining the temporal properties to both formal and simulation models, and finally the generation of the control flow graph, as shown in (Algorithm 4, lines 2–8).

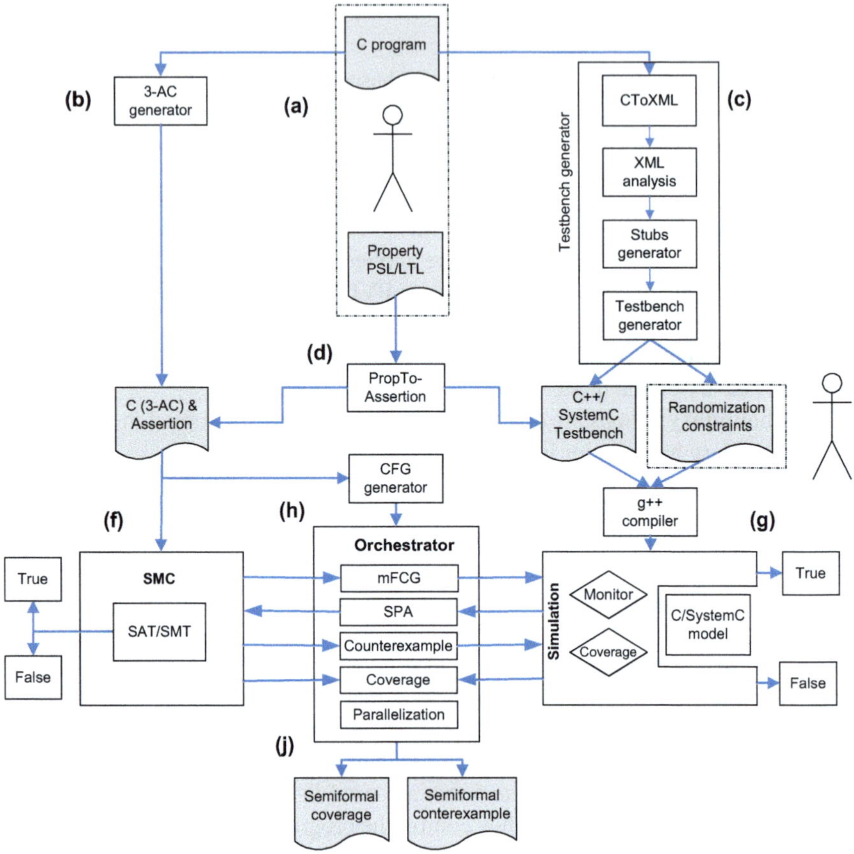

Fig. 8.2 VERIFYR overview

8.3.2.1 3-AC and FCG Generation

In order to extract the formal model we use the CIL [34] tool in the front-end to convert the C program (compatible with MISRA [35]) into three-address code (3-AC) (Fig. 8.2b). 3-AC is normally used by compilers in order to support code transformations and it is easier to handle compared to the degrees of freedom of a user implementation.

A function call graph (FCG) is generated based on [36]. We use this FCG as input to guide our Formal Exploration phase (bottom-up) verification.

8.3.2.2 Testbench Generation

For automatic testbench generation (Fig. 8.2c), we use our own XML-based approach. The objective of our testbench generator is to extract all input variables out of any

C source code and provide them in a portable overview easy to modify by hand or by using supported automatic manipulation methods. In order to reach these goals we generate a XML representation of the C code. Afterwards we analyze the corresponding XML, such as, macros, function monitoring, identification of local and input variables, value ranches of variables and loop analysis. After this step we generate the testbench using either C++ executable or SystemC model.

During the simulation we measure the coverage of the loop behavior and value ranges of all variables. The results of static XML analysis and the dynamic testbench execution are sent to the VERIFYR to enhance the automatic SPA with value ranges for variables and bounds for loop unrolling. All gathered information is presented to the user. The user can access the testbench XML description to update or manipulate the behavior.

8.3.2.3 SystemC Model

The derived simulation model is automatically generated using no abstractions. The derived model consists of one SystemC class (`ESW_SC`) mapped to a corresponding C program. The `main` function in C is converted into a SystemC process (`SC_THREAD`). Since software itself does not have any clock information, we propose a new timing reference using a program counter event (`esw_pc_event`) [37]. Additionally the `wait();` statement is necessary to suspend the SystemC process. The program counter event will be notified after every statement and will be responsible to trigger the SCTC.

The automatically generated testbench includes all input variables and it is possible to choose between different randomization strategies like constrained randomization and different random distributions, supported by the SystemC Verification Library (SCV) [38].

8.3.2.4 Temporal Properties Definition

The C language does not support any means to check temporal properties in software modules during the simulation. Therefore, we use the existing SCTC, which is a hardware oriented temporal checker based on SystemC. SCTC supports specification of properties either in PSL (Property Specification Language), LTL or FLTL (Finite Linear time Temporal Logic) [32], an extension to LTL with time bounds on temporal operators. SCTC has a synthesis engine which converts the plain text property specification into a format that can be executed during system monitoring. We translate the property to Accept-Reject automata (AR) (Fig. 8.2d) in the form of an Intermediate Language (IL) and later to a monitor in SystemC. The AR can detect validation (i.e., `True`) or violation (i.e., `False`) of properties (Fig. 8.2g) on finite system traces, or they stay in a pending state if no decision can be made yet.

For the software model checkers, we include the user-defined properties into the C code translating the LTL-style properties into assert/assume statements based on [39] (Fig. 8.2d).

8.3.3 Orchestrator

The orchestrator has as main function the coordination of the interaction between the assertion-based (i.e., simulation) and formal verification engines (Fig. 8.2h). Concerning that each SMC has their pros and cons, the formal verification is performed by the available state-of-the-art SAT/SMT based model checkers (e.g., CBMC, ESBMC). The simulation is performed by the SystemC kernel.

Additionally, the orchestrator collects the verification status of the software model checkers in order to "mark" the FCG. The mFCG is passed to the simulation engine to start the simulation process to determine values to the function parameters. Also, SPA is performed in order to identify the most important function parameter. The marked C function is updated with the static parameters and the SMC is executed in order to verify the properties. If a counterexample occurs, it will be used to guide the test vector randomization. Additionally, the orchestrator is responsible to collect the coverage data in order to determine the verification quality.

Finally, the orchestrator can distribute the computation of every function to a different verification instance of the supported SMCs (Fig. 8.3). The default distribution heuristic is a "try-all" approach, which means that all functions are checked with all supported SMCs. Furthermore, the user can orchestrate the distribution (e.g., in a cluster) of the functions manually and choose between the different SMCs by using a graphical user interface (GUI).

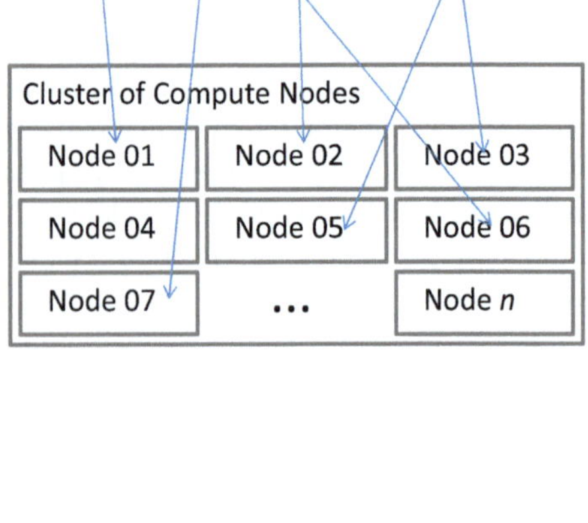

Fig. 8.3 Verification process distribution

8.3.4 Coverage

Our hybrid verification approach combines simulation-based and formal verification approaches. However, techniques to measure the achieved verification improvement have been proposed either to simulation-based or to formal verification approaches. Coverage determination for semiformal (hybrid) verification approaches is still in its infancy. For this work (Fig. 8.2j) we used a specification-based coverage metric to quantify the achieved verification improvement of hybrid software verification. Our semiformal coverage metric is based on "property coverage," which determines the total number of properties from a set of properties that were evaluated by both simulation-based or formal verification engines. Additionally, the simulation part is monitored using Gcov [40] in order to measure further implementation-based coverage inputs (e.g., line coverage, branch coverage). It is also important to point out, that due to the use of the simulation in our hybrid verification approach we still might not cover 100% of the state space, as in formal verification, as shown in Fig. 8.1b5).

8.3.5 Technical Details

The main objective of this new approach is to provide a scalable and extendable hybrid verification service. We have implemented our new approach as a verification platform called VERIFYR, which can verify embedded software in a distributed and hybrid way. To make use of the advantage of several compute nodes we have to split the whole verification process into multiple verification jobs. Furthermore, VERIFYR is platform independent and extendable by using a standard communication protocol to exchange information. The VERIFYR framework provides a service to verify a given source code written in C language. It consists of a collection of formal verification tools (such as CBMC and ESBMC), simulation tools (e.g., SCTC), and a communication gateway in order to invoke verification commands and to exchange status information of the hybrid verification process. These commands are passed to the orchestrator using the simple object access protocol (SOAP) over HTTP respectively HTTPS as shown in Fig. 8.4. The whole set of the SOAP calls are stored in the web service description language (WSDL) file for the verification service. The client application passes the SOAP document including the name of the command and its parameters such as function name, verification information and authorization credentials. As shown in Fig. 8.5 the verification clients have to send their verification requests to a super node (orchestrator). The super node distributes the requests to different verification servers. At the moment VERIFYR supports multicore compute nodes and clusters. It is possible to setup any number of verification nodes to reach the desired scalability.

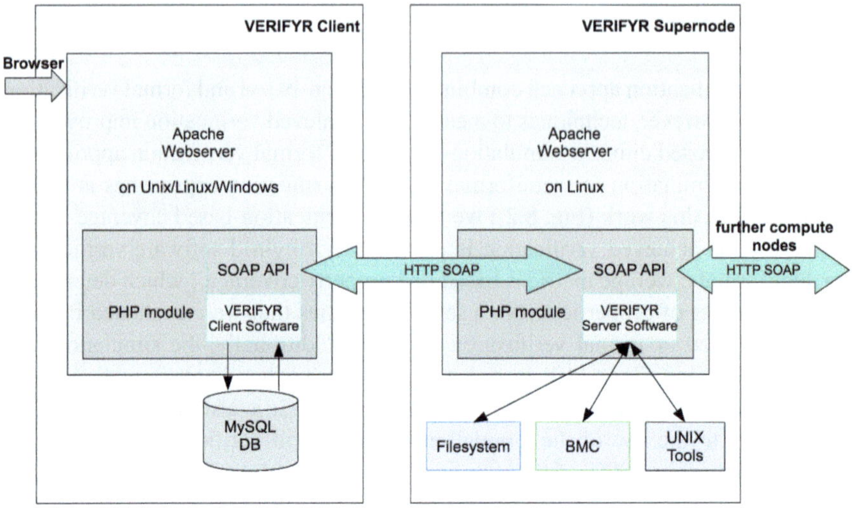

Fig. 8.4 VERIFYR client and server overview

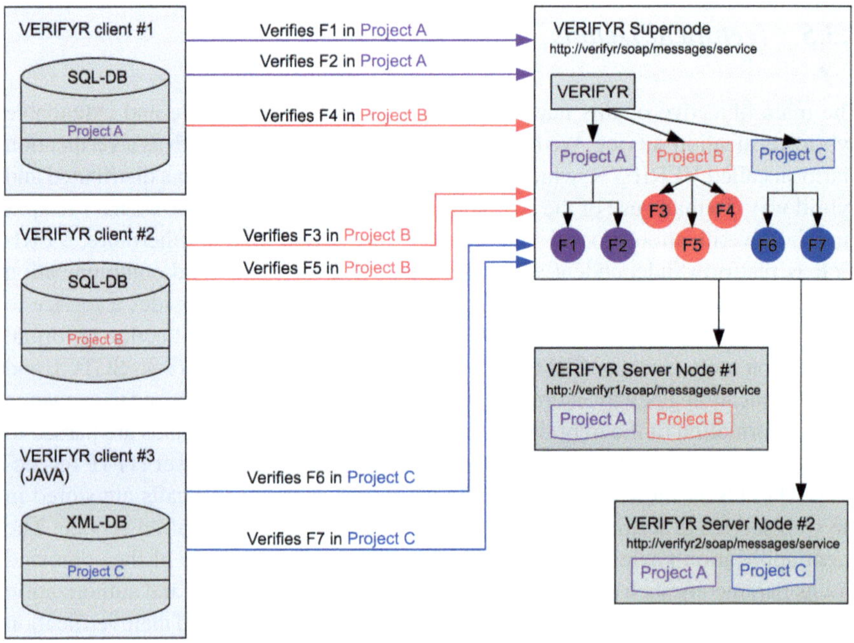

Fig. 8.5 Verification process of different clients on different servers

8.4 Results and Discussion

8.4.1 Testing Environment

We performed two sets of experiments based on two different case studies (cf., Sects. 8.4.2 and 8.4.4) conducted on a cluster with one Intel® Core™ 2 Quad CPU Q9650 @ 3.00 GHz and two Intel® Core™ 2 Duo CPU E8400 @ 3.00 GHz all with 8 GB RAM and Linux OS. The first set of experiments represents the results of the SPA heuristic based on Motorolas Benchmark Suite [41] and the verification results of our new hybrid verification methodology (VERIFYR) using this new heuristic. The second set of experiments represents the results of the SPA heuristic based on EEPROM emulation software from NEC Electronics and the verification results of the hybrid verification methodology (VERIFYR) using this new heuristic.

The scores were set according to the rules given in Sect. 8.3.1.3. The empirically gained scores are -200 points for dead parameter, -20 points for return parameters, 1025 points for conditional parameters and 2050 points for loop parameters. Then two adjustments had to be made. The first adjustment was the score of conditional parameters. Those can easily be set too low. Setting the score too low leads to a wrong ranking compared to parameters that are often used, but not as conditional or loop parameters. The actual number 1025 is an empirical choice based on the case studies. The second adjustment is the score for loop parameters. This score is reflecting used the coding style. Using many and long conditional code blocks the score for a loop decreases, while using wide loops or conditional constructions with else case the score increases. For the Motorola Powerstone Benchmark Suite the score twice the score of conditional parameters showed to be fitting. Adjusting the scores slightly will have only a small effect, but it may swap parameters that are close.

8.4.2 Motorola Powerstone Benchmark Suite

For our first case study we used Motorola's Powerstone Benchmark Suite [41] and tried to verify the built-in properties (e.g., division-by-zero) from CBMC and ESBMC. To exemplify these results the `search_dict` function from the Motorola Powerstone Benchmark module V.42 was used. The function has two parameters `string` and `data`. With these two parameters in the parameter list the algorithm proceeds through the function body. Table 8.1 shows the result for each statement. That would be 2 points for `data` and `string` and 7 points for `kid`. The score of each parameter is summed up including the appearance points. The resulting ranking is 1007 points for `data`, 3077 points for `string` and 2057 points for `kid`. However, as `kid` is inherited by `string` the score is combined to a final result of 1007 points for `data` and 5134 points for `string`.

As the score represents the impact of each parameter, it can be expected that the `string` parameter has a much bigger impact than the `data` parameter. To test the

Table 8.1 Statement scoring

Line: statement	Parameter	Points	Reason
12: if (!string)	String	1025	Switch statement
13: return (data+3);	Data	−20	Return statement
14: for (kid = dict[string].kids; …	String	2050	Loop statement, also introduces new parameter "kid"
14: kid; …	Kid	0	Not a statement, not a loop statement as the loop defines "kid"
14: kid = dict[kid].sibling)	Kid	0	Not a statement still part of the "for" instruction
15: if (kid != last && …	Kid	1025	Switch statement
15: dict[kid].data == data)	Kid	1025	Switch statement actually it is the same switch statement
15: dict[kid].data == data)	Data	1025	Switch statement two parameter, scored twice
16: return (kid);	Kid	0	One point for use not a return statement because "kid" is inherited

impact of the ranking, the function has been verified using CBMC with an unwinding option of 20 and again for each parameter. Using SPA on the parameter `data` does not change that result. The verification run resulted in about 5 s and with a memory usage of up to 175 MB. Using SPA on the parameter `string` results in a runtime of 4 s and a maximum memory usage of 65 MB. So the memory usage has been more than halved and the runtime reduced when SPA is used on the parameter the heuristic suggests. This experiment shows that SPA on the parameter improves the memory usage and the runtime. In order to show the power of SPA in more detail the function `memcpy` of the V.42 module is a good example. This function has three parameters with a scoring printed in Table 8.2. Using CBMC without unwinding this function needs more than 3 GB of memory, which leads to an out-of-memory exception in the used test environment.

Using the ranking provided by this heuristic the first parameter is a dead parameter, so applying SPA on it should lead to no further information. Applying SPA to the first parameter leads indeed to an out-of-time exception after one hour of runtime. The second parameter has a low impact on the model size. Applying SPA on the second parameter leads to another out-of-memory exception. The final parameter with the highest score has the highest impact on the model size. After applying SPA

Table 8.2 SPA results for V.42

Function	Parameter	Score	CPU[a]	Mem[b]	Vmem[b]	Comment
memcpy			458,318	276,558	2935,808	MO[c]
	*void *d*	−218	3599,565	107,089	35,652	MO[c]
	*void *s*	2	120,324	101,972	2931,712	MO[c]
	long t	1029	6,464	0,197	59,488	51[d]
strncmp			212,686	185,147	2928,64	MO[c]
	*char *s1*	2052	233,442	201,603	2939,904	MO[c]
	*char *s2*	2052	244,351	210,341	2921,472	MO[c]
	long n	8207	1,503	0,013	35,668	10704[d]

[a] seconds of runtime
[b] megabyte (virtual) memory used
[c] memory out
[d] number of clauses, all results are retrieved using CBMC with no unwind bound

on that parameter CBMC, returns "verification failed." The second experiment is the `strncmp` function of the V.42 module. This function has three parameters with the scoring shown in Table 8.2. Unlike the `memcpy` function the scoring of two parameters are close by. This suggests similar results when using SPA on either of them. Table 8.2 shows that this assumption is correct in this case. The third parameter with the highest score indeed has the highest impact on the model size and leads to a final result.

8.4.3 Verification Results Using VERIFYR

We combined the new SPA heuristic with the VERIFYR platform. We focused our interests on Modem Encoding/Decoding (*v42.c*). In total, the whole code comprises approximately 2,700 lines of C code and 12 functions. We tried to verify the built-in properties (e.g., division-by-zero, array out of bounds) from CBMC and ESBMC. It was not possible to verify the whole program using one of the above-mentioned SMCs with a *unwinding* parameter (bound) bigger than 4. For every function we used a different instance of CBMC or ESBMC in parallel. The results are shown in Table 8.3. Based on this Formal Exploration analysis, we switched to our top-down verification phase triggered by the simulation tool. At every entry point (POI), SCTC exchanges the actual variable assignment with the orchestrator, which uses this information to create temporary versions of the source code of the function under test with static assigned variables. Table 8.3 shows the comparison between CBMC (SAT), ESBMC, and our VERIFYR platform. The used symbols are P (passed), F (failed), MO (out of memory), TO (time out, 90 min), and PH (passed using hybrid methodology). PH means that it was possible to verify this function with our hybrid methodology using simulation to support formal verification with static parameter assignment. This table shows that VERIFYR presented the same valid results as CBMC (SAT)

Table 8.3 Verification results v42.c

Function	CBMC (SAT)		ESBMC		VERIFYR	
	Result	Time (s)	Result	Time (s)	Result	Time (s)
Leaves						
putcode	P	2	P	2	P	2
getdata	P	2	P	2	P	2
add_dict	MO	135	MO	155	PH	535
init_dict	MO	152	P	40	P	40
search_dict	MO	161	MO	234	PH	535
putdata	P	1	P	1	P	1
getcode	P	1	P	1	P	1
puts	MO	163	MO	134	PH	535
Parents level 1						
checksize_dict	TO		TO		PH	535
encode	MO	354	MO	289	PH	2
decode	P	1	P	1	P	1
ALL						
main	MO	351	MO	274	PH	535

P (passed), F (failed), MO (out of memory),
TO (time out, 90 min) and PH (passed using hybrid methodology)

and ESBMC, and no MO or TO has occurred. Furthermore, the Table 8.3 presents the verification time in seconds in order to reach P, MO, or PH results. The time for PH consist of the time for the simulation runs plus formal verification using static parameter assignment. We have used 1000 simulation runs. In total, 20 properties were evaluated by both simulation and formal verification. All tested properties were safe, that is, a property coverage of 100%.

Overall, we have simulated the whole modem encoding/decoding software using our automatically generated testbench and beyond that we are able to verify 6 out of 12 observed functions using formal verification and the 6 remaining with hybrid verification. However, VERIFYR outperforms the single state-of-the-art tools in complex cases where they are not capable to reach a final verification result.

8.4.4 EEPROM Emulation Software from NEC Electronics

Our second case study is an automotive EEPROM Emulation software from NEC Electronics [42], which emulates the read and write requests to a nonvolatile memory. This embedded software contains both hardware-independent and hardware-dependent layers. Therefore, this system is a suitable automotive industrial application to evaluate the developed methodologies with respect to both abstraction layers. The code used is property of NEC Electronics (Europe) GmbH, embedded and marked confidential. Therefore, the details of the implementation are not discussed. The EEPROM emulation software uses a layered approach divided into two parts: the Data Flash Access layer (DFALib) and the EEPROM Emulation layer (EEELib). The Data Flash Access layer is a hardware-dependent software layer that provides an easy-to-use interface for the FLASH hardware. The EEPROM Emulation layer is a hardware-independent software layer and provides a set of higher level operations for the application level. These operations include: `Format`, `Prepare`, `Read`, `Write`, `Refresh`, `Startup1` and `Startup2`. In total, the whole EEPROM emulation code comprises approximately 8,500 lines of C code and 81 functions. We extracted from the NEC specification manual two property sets (LTL standard). Each property in the EEELib set describes the basic functionality on each EEELibs operation (i.e., read, write, etc.). A sample of our LTL properties is as follows:

$$\mathsf{F}\ (\text{Read} \rightarrow \text{X F(EEE\_OK}|| \ldots)) \quad \text{(A)}$$

The property represents the calling operations in the EEELib library (e.g., Read) and several return values (e.g., EEE_OK) that may be received. For CBMC we translated the LTL properties to assert/assume style properties based on [39]. For the SPA heuristic the same scoring as in the Motorola Powerstone Benchmark Suite was used. The verification was done on the same computer as the previous testing and the verification runs were unbounded. We present three functions to provide evidence that the concept of this heuristic is valid and the scoring is balanced. In Table 8.4 the measured results are shown.

The function `DFA_Wr` is successfully verified using SPA on the `length` parameter. This result is suggested by the heuristic. In the function `DFA_WrSec` the parameter `val` has the highest score. And the function also finishes using SPA on that parameter. Unlike in the two other functions the function `DFALib_SetWr` is valid from the beginning. CBMC verifies the function in half a second using 1341 clauses. Still using SPA shows that if the score of the parameters increase then the number of clauses generated and proven by CBMC decreases. This shows that the score is representing the complexity of parameters concerning the resulting state space. Unbounded model checking can be restricted in order to gain a partial result. The case studies above show that the increased complexity of software can be handled using SPA.

We have selected for both EEELib and DFALib (hardware-dependent) two leaf functions and two corresponding parent functions in relation to the corresponding FCG. We have renamed the selected functions for convenience. Table 8.5 shows that VERIFYR presented the same valid results as CBMC (SAT) and ESBMC, and no MO or TO has occurred. In total, 40 properties were evaluated by both simulation and formal verification, which corresponds five properties for each of the eight functions. All tested properties were safe, that is, a property coverage of 100%.

Table 8.4 SPA results for NEC

Function	Parameter	Score	CPU[a]	Mem.[b]	Vmem.[b]	Comment
DFA_Wr			127,576	107,435	2917,376	MO[c]
	*void *addSrc*	15	127,964	106,242	2929,664	MO[c]
	*void *addDest*	15	138,541	116,533	2934,784	MO[c]
	u32 length	3093	0,534	0	0	VS[d]
DFA_WrSec			129,523	109,166	2939,904	MO
	*u08 volatile *sec*	−199	129,826	106,599	2934,784	MO[c]
	*u08 volatile *dest*	829	133,273	111,806	2936,832	MO[c]
	u08 mask	1852	123,984	105,334	2922,496	MO[c]
	u08 val	4115	0,521	0,002	21,141	51[e]
DFA_SetWr			0,552	0,003	21,148	1341[e]
	*u32 *pWrite-data*	19	0,541	0	0	1031[e]
	u32 cnt	1853	0,5	0	0	29[e]

[a]seconds of runtime
[b]megabyte (virtual) memory used
[c]memory out
[d]verification successful
[e]number of clauses, all results are retrieved using CBMC with no unwind bound

Table 8.5 Verification Results NEC

Function	CBMC (SAT)		ESBMC		VERIFYR	
	Result	Time (s)	Result	Time (s)	Result	Time (s)
EEELib						
Eee_Leaf01	P	1	P	1	P	1
Eee_Leaf02	P	1	P	1	P	1
Eee_Parent01	MO	231	MO	174	PH	1840
Eee_Parent02	MO	110	MO	119	PH	1840
DFALib						
DFA_Leaf01	P	1	P	1	P	1
DFA_Leaf02	MO	109	MO	90	PH	1840
DFA_Parent01	MO	112	MO	92	PH	1840
DFA_Parent02	MO	125	MO	100	PH	1840

P (passed), F (failed), MO (out of memory),
TO (time out, 90 min) and PH (passed using hybrid methodology)

Overall, when we look at the results, we have simulated the whole NEC software using our generated testbench and beyond that we were able to verify 3 out of 8 observed functions using formal verification and the remaining using hybrid verification. VERIFYR outperforms the state-of-the-art tools in this complex application where they are not able to reach a final verification result for all functions.

8.5 Conclusion and Future Work

We have presented our scalable and extendable hybrid verification approach for embedded software. We have described our new semiformal verification methodology and have pointed out the advantages. Furthermore we have shown our new SPA heuristic, which shows promising results on the Motorola Powerstone Benchmarks Suite and on the EEPROM emulation software from NEC Electronics. SPA is an automated process that optimizes the interaction between bounded model checking and simulation for semiformal verification approaches. It is possible to use different strategies for the whole or parts of the verification process. We start with the formal phase and end up with hybrid verification based on simulation and formal verification. During the formal exploration phase the SMC tries to verify all possible functions under test based on a FCG until a time bound or memory limit has been reached. The FCG is marked to indicate the Points-of-Interest. Then, we start with simulation and whenever one of the POIs is reached, the orchestrator generates a temporary version of the function under test with initialized/pre-defined variables in order to shrink the state space of the formal verification. Our results show that the whole approach is best suited for complex embedded C software with and without hardware dependencies. It scales better than standalone software model checkers and reaches deep state spaces. Furthermore, our approach can be easily integrated in a complex software development process. Currently, we are working on assessing the scores automatically and on quality metrics for hybrid verification.

Acknowledgements The authors would like to thank Edgar Auerswald, Patrick Koecher and Sebastian Welsch for supporting the development of the VERIFYR platform.

References

1. Jerraya AA, Yoo S, Verkest D, Wehn N (2003) Embedded software for SoC. Kluwer Academic Publishers, Norwell, MA, USA
2. Beyer D, Henzinger TA, Jhala R, Majumdar R (2007) The software model checker BLAST: applications to software engineering. Int J Softw Tools Technol Trans
3. Behrend J, Lettnin D, Heckler P, Ruf J, Kropf T, Rosenstiel W (2011) Scalable hybrid verification for embedded software. In: DATE '11: proceedings of the conference on design, automation and test in Europe, pp 1–6
4. Barrett C, Sebastiani R, Seshia SA, Tinelli C (2009) Satisfiability modulo theories. Frontiers in artificial intelligence and applications, Chap 26, vol 185. IOS Press, pp 825–885

5. Clarke E, Kroening D, Lerda F (2004) A tool for checking ANSI-C programs. In: Tools and algorithms for the construction and analysis of systems. Springer, pp 168–176
6. Kroening D (2009) Bounded model checking for ANSI-C. http://www.cprover.org/cbmc/
7. Biere A, Cimatti A, Clarke EM, Strichman O, Zhu Y (2003) Bounded model checking. In: Zelkowitz M (ed) Highly dependable software. Advances in computers, vol 58. Academic Press
8. Cordeiro L, Fischer B, Marques-Silva J (2009) SMT-based bounded model checking for embedded ANSI-C software. In: ASE'09: proceedings of the 2009 IEEE/ACM international conference on automated software engineering. IEEE Computer Society, Washington, DC, pp 137–148
9. Ball T, Majumdar R, Millstein T, Rajamani SK (2001) Automatic predicate abstraction of C programs. SIGPLAN Not 36:203–213
10. Flanagan C, Qadeer S (2002) Predicate abstraction for software verification. SIGPLAN Not 37:191–202
11. Clarke E, Grumberg O, Jha S, Lu Y, Veith H (2003) Counterexample-guided abstraction refinement for symbolic model checking. J ACM 50:752–794
12. Clarke E, Grumberg O, Long D (1994) Model checking and abstraction. ACM Trans Prog Lang syst 16(5):1512–1542
13. Henzinger TA, Jhala R, Majumdar R (2005) The BLAST software verification system. Model Checking Softw 3639:25–26
14. Clarke E, Kroening D, Sharygina N, Yorav K (2005) SATABS: SAT-based predicate abstraction for ANSI-C. In: TACAS, vol 3440. Springer, pp 570–574
15. Gorai S, Biswas S, Bhatia L, Tiwari P, Mitra RS (2006) Directed-simulation assisted formal verification of serial protocol and bridge. In: DAC '06: proceedings of the 43rd annual design automation conference. ACM, New York, pp 731–736
16. Nanshi K, Somenzi F (2006) Guiding simulation with increasingly refined abstract traces. In: DAC '06: proceedings of the 43rd annual design automation conference. ACM, New York, pp 737–742
17. Di Guglielmo G, Fummi F, Pravadelli G, Soffia S, Roveri M (2010) Semi-formal functional verification by EFSM traversing via NuSMV. In: 2010 IEEE international High level design validation and test workshop (HLDVT), pp 58–65
18. Edwards SA, Ma T, Damiano R (2001) Using a hardware model checker to verify software. In: proceedings of the 4th international conference on ASIC (ASICON)
19. Lettnin D, Nalla PK, Behrend J, Ruf J, Gerlach J, Kropf T, Rosenstiel W, Schönknecht V, Reitemeyer S (2009) Semiformal verification of temporal properties in automotive hardware dependent software. In: DATE'09: proceedings of the conference on design, automation and test in Europe, pp 1214–1217
20. Ruf J, Peranandam PM, Kropf T, Rosenstiel W (2003) Bounded property checking with symbolic simulation. In: FDL
21. Cordeiro L, Fischer B, Chen H, Marques-Silva J (2009) Semiformal verification of embedded software in medical devices considering stringent hardware constraints. In: Second international conference on embedded software and systems, pp 396–403
22. Godefroid P, Klarlund N, Sen K (2005) Dart: directed automated random testing. SIGPLAN Not 40(6):213–223. http://doi.acm.org/10.1145/1064978.1065036
23. Cadar C, Ganesh V, Pawlowski PM, Dill DL, Engler DR (2006) Exe: automatically generating inputs of death. In: Proceedings of the 13th ACM conference on computer and communications security. CCS'06. ACM, New York, pp 322–335. http://doi.acm.org/10.1145/1180405.1180445
24. Sen K, Marinov D, Agha G (2005) CUTE: a concolic unit testing engine for C. SIGSOFT Softw Eng Notes 30(5):263–272. http://doi.acm.org/10.1145/1095430.1081750
25. Di Guglielmo G, Fujita M, Fummi F, Pravadelli G, Soffia S (2011) EFSM-based model-driven approach to concolic testing of system-level design. In: 2011 9th IEEE/ACM international conference on formal methods and models for codesign (MEMOCODE), pp 201–209

26. Cadar C, Dunbar D, Engler D (2008) KLEE: unassisted and automatic generation of high-coverage tests for complex systems programs. In: Proceedings of the 8th USENIX conference on operating systems design and implementation. OSDI'08. USENIX Association, Berkeley, pp 209–224
27. Lattner C, Adve V (2005) The llvm compiler framework and infrastructure tutorial. In: Eigenmann R, Li Z, Midkiff S (eds) Languages and compilers for high performance computing. Lecture notes in computer science, vol 3602. Springer, Berlin, pp 15–16
28. Tillmann N, De Halleux J (2008) Pex: white box test generation for .net. In: Proceedings of the 2nd international conference on tests and proofs. TAP'08. Springer, Heidelberg, pp 134–153. http://dl.acm.org/citation.cfm?id=1792786.1792798
29. Cuoq P, Kirchner F, Kosmatov N, Prevosto V, Signoles J, Yakobowski B (2012) Frama-C: a software analysis perspective. In: Proceedings of the 10th international conference on software engineering and formal methods. SEFM'12. Springer, Heidelberg, pp 233–247
30. Correnson L, Signoles J (2012) Combining analyses for C program verification. In: Stoelinga M, Pinger R (eds) Formal methods for industrial critical systems. Lecture notes in computer science, vol 7437. Springer, Berlin, pp 108–130
31. Kirchner F, Kosmatov N, Prevosto V, Signoles J, Yakobowski B (2015) Frama-C: a software analysis perspective. Formal Aspects Comput:1–37
32. Weiss RJ, Ruf J, Kropf T, Rosenstiel W (2005) Efficient and customizable integration of temporal properties into SystemC. In: Forum on specification & design languages (FDL), pp 271–282
33. Clarke E, Grumberg O, Hamaguchi K (1994) Another look at LTL model checking. In: Dill DL (ed) Conference on computer aided verification (CAV). Lecture notes in computer science, vol 818. Springer, Stanford, pp 415–427
34. Necula GC, McPeak S, Rahul SP, Weimer W (2002) CIL: intermediate language and tools for analysis and transformation of C programs. In: Computational complexity, pp 213–228
35. MISRA (2000) MISRA—the motor industry software reliability association. http://www.misra.org.uk/
36. Shea R (2009) Call graph visualization for C and TinyOS programs. In: Department of computer science school of engineering UCLA. http://www.ambleramble.org/callgraph/index.html
37. Lettnin D, Nalla PK, Ruf J, Kropf T, Rosenstiel W, Kirsten T, Schönknecht V, Reitemeyer S (2008) Verification of temporal properties in automotive embedded software. In: DATE'08: proceedings of the conference on design, automation and test in Europe. ACM, New York, pp 164–169
38. Open SystemC Initiative (2003) SystemC verification standard library 1.0p users manual
39. Clarke E, Kroening D, Yorav K (2003) Behavioral consistency of C and verilog programs using bounded model checking. In: DAC'03: proceedings of the 40th annual design automation conference. ACM, New York, pp 368–371
40. GNU (2010) Gcov coverage. http://gcc.gnu.org/onlinedocs/gcc/Gcov.html
41. Malik A, Moyer B, Cermak D (2000) The M'CORE (TM) M340 unified cache architecture. In: Proceedings of the 2000 international conference on computer design, pp 577–580
42. NEC NEC Electronics (Europe) GmbH. http://www.eu.necel.com/

Index

D. Lettnin and M. Winterholer (eds.), *Embedded Software Verification and Debugging*, Embedded Systems, DOI 10.1007/978-1-4614-2266-2

MIX
Papier aus verantwortungsvollen Quellen
Paper from responsible sources
FSC® C105338

If you have any concerns about our products,
you can contact us on
ProductSafety@springernature.com

In case Publisher is established outside the EU,
the EU authorized representative is:
Springer Nature Customer Service Center GmbH
Europaplatz 3, 69115 Heidelberg, Germany

Printed by Libri Plureos GmbH
in Hamburg, Germany